# Barn Owls

With heart-shaped face, buff back and wings, and pure white underparts, the barn owl is a distinctive and much-loved bird that has fascinated people from many cultures throughout history. How did the barn owl colonize the world? What adaptations have made this bird so successful? How is the increasing impact of human disturbance affecting these animals? Answering these questions and more, Alexandre Roulin brings together the main global perspectives on the evolution, ecology and behaviour of the barn owl and its relatives, discussing topics such as high reproductive potential, physiology, social and family interactions, pronounced colour variation and global distribution. Accessible and beautifully illustrated, this definitive volume on the barn owl is for researchers, professionals and graduate students in ornithology, animal behaviour, ecology, conservation biology and evolutionary biology, and will also appeal to amateur ornithologists and nature lovers.

**Alexandre Roulin** is a professor of biology at the University of Lausanne, Switzerland. For three decades, he has studied barn owls to answer evolutionary and ecological questions. His main scientific interests are the adaptive function of melanin-based colouration and negotiation processes taking place in animal societies. Since 2009, he has actively participated in a project that harnesses ecology and farming to promote reconciliation between Israeli, Palestinian and Jordanian communities through nature-based solutions. He strives to reconcile humans with nature and supports interdisciplinary approaches promoting peace and respect for the environment.

In memory of the late Martin Epars

Adoption

Double breeding  Synanthrope

## Colour morphs

Reproductive potential

# Cosmopolitan

Hatching asynchrony  **Sibling negotiation**

Pellets

Offspring desertion

Food sharing

Fly silently  Food stores

Biological pest control agent

Hearing capacity

# Barn Owls

## Evolution and Ecology

ALEXANDRE ROULIN

*University of Lausanne, Switzerland*

Artwork by Laurent Willenegger

CAMBRIDGE
UNIVERSITY PRESS

# CAMBRIDGE
## UNIVERSITY PRESS

University Printing House, Cambridge CB2 8BS, United Kingdom

One Liberty Plaza, 20th Floor, New York, NY 10006, USA

477 Williamstown Road, Port Melbourne, VIC 3207, Australia

314–321, 3rd Floor, Plot 3, Splendor Forum, Jasola District Centre, New Delhi – 110025, India

79 Anson Road, #06–04/06, Singapore 079906

Cambridge University Press is part of the University of Cambridge.

It furthers the University's mission by disseminating knowledge in the pursuit of education, learning and research at the highest international levels of excellence.

www.cambridge.org
Information on this title: www.cambridge.org/9781107165755
DOI: 10.1017/9781316694114

© Cambridge University Press 2020

First published 2020

Printed in Singapore by Markono Print Media Pte Ltd

*A catalogue record for this publication is available from the British Library.*

*Library of Congress Cataloging-in-Publication Data*
Names: Roulin, Alexandre, 1968–author.
Title: Barn owls : evolution and ecology / Alexandre Roulin, University of
    Lausanne, Switzerland ; artwork by Laurent Willenegger.
Description: Cambridge, United Kingdom ; New York, NY : Cambridge University
    Press, 2019. | Includes bibliographical references and index.
Identifiers: LCCN 2019008026 | ISBN 9781107165755 (hardback)
Subjects: LCSH: Barn owl. | Barn owl–Evolution. | Barn owl–Ecology. | Barn owl–Behavior.
Classification: LCC QL696.S85 R68 2019 | DDC 598.9/7–dc23
LC record available at https://lccn.loc.gov/2019008026

ISBN 978-1-107-16575-5 Hardback

# CONTENTS

*Foreword by Erkki Korpimäki*      *Page* xiii
*Acknowledgements*      xv
*Why this book?*      xvii

## 1 Introduction    1
1.1 Why is the barn owl so interesting?    2
1.2 Why study barn owls instead of laboratory mice?    6
1.3 The raw data    8
1.4 Evolution of the Tytonidae    10
1.5 Why cosmopolitan?    18
1.6 Why do barn owls live so close to humans?    23

## 2 Conservation    27
2.1 Why protect barn owls?    28
2.2 Ethics    32
2.3 Decreases in barn owl populations    36
2.4 Pollution    39
2.5 What can we do to protect barn owls?    44

## 3 Parasites and predators    53
3.1 Endoparasites    54
3.2 Ectoparasites    58
3.3 Predators and anti-predator behaviour    62

## 4 Physiology in an ecological context    65
4.1 Hearing capacity    66
4.2 Visual capacity    68
4.3 Daily food requirements    71
4.4 Pellet production    74
4.5 Capacity to withstand cold weather    76

## 5 Morphology in an ecological context    81
5.1 Body size    82
5.2 Reversed sexual size dimorphism    88

## 6 Daily life: hunting, feeding and sleeping    91
6.1 A nocturnal hunter    92
6.2 Roosting    95
6.3 Home range    100
6.4 Flight mechanics    106

|       | 6.5  Hunting methods                            | 112 |
|       | 6.6  Prey selection                             | 116 |
|       | 6.7  Diet                                       | 118 |

| **7**  | **Sexual behaviour**                           | **127** |
|       | 7.1  Courtship and copulation                   | 128 |
|       | 7.2  Mating system                              | 132 |
|       | 7.3  Fidelity and divorce                       | 138 |

| **8**  | **Reproduction**                               | **143** |
|       | 8.1  Nest sites                                 | 144 |
|       | 8.2  Interspecific competition over nest sites  | 150 |
|       | 8.3  Reproductive season                        | 152 |
|       | 8.4  Egg formation                              | 155 |
|       | 8.5  Clutch size                                | 158 |
|       | 8.6  Incubation                                 | 161 |
|       | 8.7  Hatching                                   | 163 |
|       | 8.8  Brood size                                 | 167 |
|       | 8.9  Nestling growth                            | 173 |
|       | 8.10 Second and third annual broods             | 176 |
|       | 8.11 Offspring desertion                        | 179 |

| **9**  | **Parental care**                              | **183** |
|       | 9.1  Parental foraging                          | 184 |
|       | 9.2  Parental behaviour at the nest             | 188 |
|       | 9.3  Adult body mass                            | 191 |
|       | 9.4  Food stores                                | 194 |
|       | 9.5  Adoption                                   | 197 |

| **10** | **Sibling interactions**                       | **199** |
|       | 10.1 Timing of nestling activities              | 200 |
|       | 10.2 Sibling negotiation                        | 203 |
|       | 10.3 Begging behaviour                          | 210 |
|       | 10.4 Stealing food from siblings                | 215 |
|       | 10.5 Food sharing among siblings                | 217 |
|       | 10.6 Mutual preening                            | 219 |
|       | 10.7 Social huddling                            | 221 |

| **11** | **Demography**                                 | **223** |
|       | 11.1 Natal dispersal                            | 224 |
|       | 11.2 Breeding dispersal and migration           | 228 |
|       | 11.3 Survival prospects                         | 230 |
|       | 11.4 Population dynamics                         | 235 |

## 12 Plumage polymorphism 239

12.1 Colour polymorphism 240
12.2 Genetics of plumage polymorphism 245
12.3 Sexual dimorphism in plumage traits 252
12.4 Age-related changes in plumage traits 254
12.5 Mate choice 257
12.6 Sexually antagonistic selection 260
12.7 Adaptive functions of whitish and reddish colouration 264
12.8 Adaptive functions of small and large black spots 268
12.9 Adaptive functions of few and many black spots 275
12.10 Geographic variation in plumage traits 278

## Conclusion 285

To the future 286
Species names 288

*Index* 291

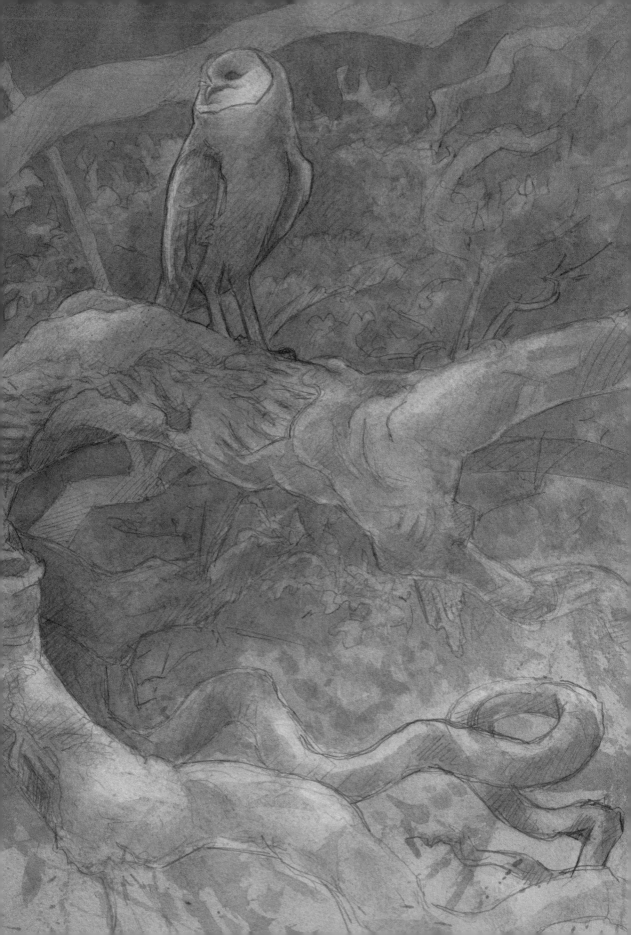

Grass owl in Taiwan, November 2014. ©Yi-Shuo Tseng

# FOREWORD

Large owls are top predators that are often portrayed as magnificent, regal and powerful, while smaller owls are considered handsome and cute. Therefore, members of the public are interested in owls not only in the wild but also as pet animals.

Both large and smaller owls capture mainly small mammals, particularly mice, rats and voles. These rodents are considered pests, because they damage crops on agricultural land and in gardens, and saplings in forestry plantations. Traditionally, several rodenticides have been used to prevent or at least diminish economic losses caused by small-rodent damage, but these often impose mortality not only on the target pest animals but also on their predators, including owls. From the 1990s onwards, a substantial body of observational and experimental studies showed that owls and other predators can limit or even control population densities of small rodents. They can thus probably be used as biological pest control agents to reduce damage from small rodents to agricultural crops and saplings. This exemplifies the extent to which owls are useful to humans and should be protected.

That we know so much about secretive owls is a credit to many dedicated fieldworkers and field scientists, along with academics. In particular, the population dynamics, demography and diet composition of owls have been the subjects of many long-term studies in temperate, boreal and arctic regions – requiring great physical effort – but less so in the tropics. Because of intensification of forest management and agricultural practices, the habitats of forest-dwelling and open-country owls – such as Tengmalm's owls, spotted owls, short-eared owls and barn owls – have been degraded over the past fifty years. In addition, climate change has imposed habitat loss and thus declines of many owl species in boreal, arctic and arid regions. As a result, populations of several owl species have declined drastically, to the point that many are currently vulnerable or even close to extinction.

Barn owls are found nearly all over the world apart from in my home country and other northern European countries. Barn owls have high reproductive potential, and their social behaviour, wide colour variation and cosmopolitan distribution offer unique systems for studies on population dynamics and evolutionary ecology. I have followed the long-term research conducted and hundreds of scientific papers published by the research team led by Professor Alexandre Roulin over the past thirty years with great interest. His team has been and still is one of the leading research groups on owls worldwide and has continuously published high-quality scientific papers in international journals. He and I have similar backgrounds in the sense that we both started as keen amateur ornithologists studying owls long before becoming academics. We have put much effort into collecting long-term observational data on owl populations over large areas, and have subsequently planned and conducted field experiments to better test the hypotheses formulated on the basis of accumulated data.

For more than thirty years, Professor Roulin's team has studied population dynamics, demography, cooperation and conflict among siblings and parents, as well as melanin-based plumage colouration and its connections with behaviour including mating systems and parental care in barn owls. As a result of his work, barn owls have become one of the most important model species in studies of how natural selection operates in the wild, by producing various colour types with behavioural and physiological differences. It is truly wonderful that Alexandre Roulin has now been able to summarize his excellent studies on barn owls and their allies in this seminal book, which is intended not only for academics but also for amateur birdwatchers, nature lovers and conservationists. I have really enjoyed reading this comprehensive and comprehensible book with its gorgeous artwork.

**Erkki Korpimäki**
Professor of Animal Ecology, University of Turku, Finland

# ACKNOWLEDGEMENTS

This book is largely based on the synthesis of thousands of publications on the barn owl and its relatives. I am greatly indebted to the Swiss Ornithological Institute (die Schweizerische Vogelwarte Sempach), which was instrumental in providing access to this knowledge-base, as well as to the numerous colleagues who sent me publications I could not obtain directly. I am also grateful to the financial support provided by the following organizations: Fondation Bataillard (through my friend Professor Daniel Chérix), Fondation Chuard Schmid of the University of Lausanne, Fonds du Dr Rub of the Faculty of Biology and Medicine of the University of Lausanne, Hilfsfonds of the Swiss Ornithological Institute, as well as the Swiss National Science Foundation and the University of Lausanne for their continuous support of my research.

I thank the following colleagues for their iterative feedback while I was writing this book: Hugh Brazier, Alexandre Chausson, Luis San José Garcia, Andrea Romano, Amélie Dreiss, David Ramsden, Jeff Martin, Rohan Bilney, Pauline Ducouret, Vera Uva, Richard Prum, Martin Paeckert, Yoav Motro, Inês Roque, Rui Lourenço, Hermann Wagner, Michael Wink, Daniel Osorio, Laura Hausmann, Franck Ruffier, David Eilam, Olivier Krüger, Thomas Bachmann, Madeleine Scriba, Peter Sunde and Res Altwegg. I would also like to thank the numerous photographers whose photos I was unfortunately not able to incorporate owing to space limitations. The maps were drawn by Alexandre Hirzel with the help of Olivier Brönnimann. This book was in part written at the Wissenschaftskolleg zu Berlin.

My biggest thanks go to Laurent Willenegger, with whom I worked for more than four years to achieve a marriage of text and illustrations. I am proud to have this book illustrated by such a master.

Finally, and most importantly, I thank the barn owl as a flagship species for peace and biodiversity.

**Alexandre Roulin**

## A NOTE ABOUT THE ILLUSTRATIONS

When I was a teenager in the 1990s, I started watching barn owls. From an early age, ringing sessions provided an ideal way to study this secretive bird. And then, in 2014, Alexandre Roulin and I started to collaborate on this book. I accompanied his team in the field to observe the barn owl again. However, most of the drawings and paintings for this book were done not in the field (something I usually do) but in the studio to produce the necessary illustrations to complement the text. This work is based on my own experience with the bird, supplemented by information gathered in books or from the internet, and by material provided by Alexandre and his team. Some of the illustrations are done in watercolour, some with a pencil, and some have been produced digitally.

I hope that you have enjoyed the drawings and paintings. If you want to know more about my work, you can contact me via my website at www.wildsideproductions.ch.

**Laurent Willenegger**

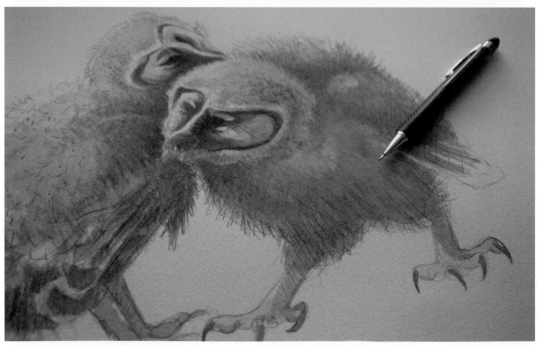

# WHY THIS BOOK?

The barn owl and its relatives (grass, masked and sooty owls) are emblematic for ornithologists and the lay public alike. Many aspects of the lifestyle and life history of the barn owl, including its high reproductive potential, complex social behaviour, pronounced colour variation and cosmopolitan distribution, make this bird a fascinating creature. It is no surprise then that the barn owl has been widely studied by scientists. A search of the *Web of Science* in 2018 revealed 1228 papers published on the genus *Tyto*, compared with 273 papers on Tengmalm's owl, 701 on the European kestrel and 3832 on the great tit, a species studied by many professional and amateur ornithologists.

Several books are dedicated to the barn owl. However, because the literature on the barn owl encompasses many topics, an updated review of worldwide knowledge about its evolutionary ecology is necessary. Although I have studied this bird for the last thirty years and have published a number of papers, this book is not just a summary of my own studies and ideas. My aim is to highlight the facts in a concise and objective way. I have been careful to avoid relying solely on literature about European or North American barn owls, as I find this approach very restrictive. For instance, the names 'barn owl' and 'effraie des clochers' ('owl of the churches') are very European-centred and do not at all characterize the Tytonidae living in Australasia or the Caribbean, where these owls often exploit forests.

This book is for anyone interested in barn, grass, masked and sooty owls, as well as in birds and nature in general. Having originally been an enthusiastic amateur ornithologist before becoming an academic, I have deliberately adopted a less formal writing style than academics are used to. I have simplified concepts without (I hope) losing their essence, to offer an opportunity to nature lovers to gain an insight into the daily life of owls as seen by academics. In an attempt to value the approach taken by people who contemplate nature from a non-research perspective, the book includes photographs obtained from a number of ornithologists, as well as drawings and paintings by an accomplished artist, Laurent Willenegger.

**Alexandre Roulin**

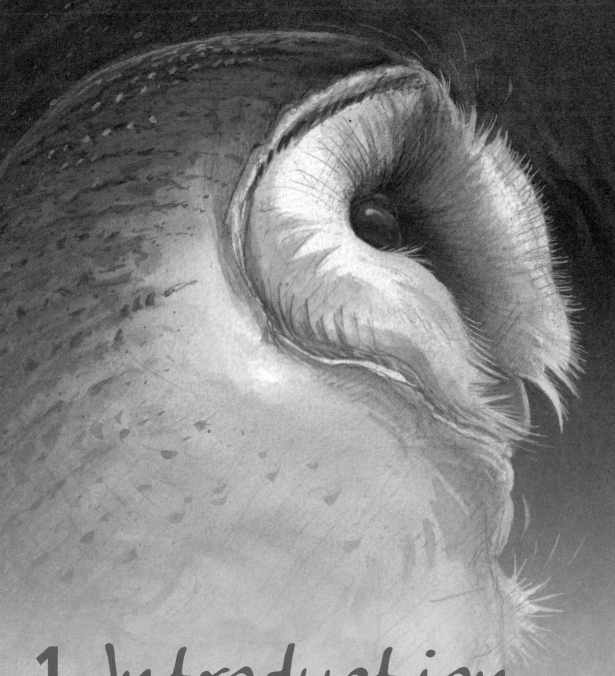

# 1 Introduction

**1.1**   Why is the barn owl so interesting?

**1.2**   Why study barn owls instead of laboratory mice?

**1.3**   The raw data

**1.4**   Evolution of the Tytonidae

**1.5**   Why cosmopolitan?

**1.6**   Why do barn owls live so close to humans?

Barn owl flying to new horizons.

## 1.1   WHY IS THE BARN OWL SO INTERESTING?

# A step above the others

**The barn owl is popular because of its cosmopolitan distribution and close association with humans. Owls have fascinated us throughout history and across many cultures. Venerated or considered as bad omens, owls feature often in myths and legends. The barn owl is a fantastic organism for investigating physiology, reproductive biology, social interactions between family members and the causes and consequences of colour variation. More to come!**

Barn owls fascinate not only the lay public but also scientists. This bird has traditionally been used to study the diet of avian predators and to teach ecology. Barn owl pellets contain extraordinarily well-preserved bones and cuticles of their prey. Among the 3696 papers reviewed for the present book, almost half (1630) were about diet analyses. But interest in the barn owl extends to many other areas as well. Compared to that of other species, the barn owl's variation in morphology, physiology and behaviour is quite remarkable. Some of the barn owl's physiological traits, such as its hearing capacity, are extremely efficient, while other traits, such as resistance to cold, are poorly developed.

   The barn owl can therefore be considered a model organism for identifying general rules that govern the evolution of biodiversity. My aim in this book is to highlight the unique characteristics of this bird in an attempt to depict general biological phenomena. These characteristics include seven key traits.

**Cosmopolitan**   Barn owls are found almost everywhere except in cold temperate and arctic regions. Only approximately twenty vertebrate species, including humans, are cosmopolitan. Studying barn owls can therefore help us understand the factors allowing some animals to thrive worldwide.

**Physiology** The barn owl is often used to study hearing capacity and the ability to fly silently. Its nocturnal habits and high reproductive rate have exerted strong selective pressure to evolve the ability to hunt very efficiently, necessitating the ability to detect prey by ear at great distances while flying stealthily. These abilities are of great interest to neurologists and to the aeronautical industry.

**Reproductive potential** Compared with other raptors, barn owls are prolific breeders. Annually they may produce up to three broods, with some containing up to twelve young. In places where conditions are favourable, they can breed throughout the year. This potential for reproduction enables an investigation into the evolutionary causes and consequences of such high reproductive rates. A large body of data can be collected from a great many descendants, which is unusual for a species of this size.

**Hatching asynchrony** The first nestling can be up to one month older than its youngest sibling. This is because the parents usually stagger the timing of the hatchings, which has the effect of generating less competition among the young for parental attention. Although hatching asynchrony is traditionally considered a way for parents to reduce brood size, a plethora of hypotheses have been presented to explain its adaptive function. Because the degree of hatching asynchrony in the barn owl is one of the most pronounced among birds, this species is a prime candidate for studying the causes and consequences of size differences between young siblings.

**Peaceful sib–sib interactions** Young siblings vocally negotiate priority of access to food resources, and nestlings feed and preen their siblings. Social interactions within barn owl nests are intricate: this provides insight into parent–offspring conflict and sibling competition and cooperation. Although, in any species, siblings are genetically related, observations and theory have predicted that family interactions tend to be conflictual rather than harmonious. The finding that young barn owls peacefully share resources suggests that sibling cooperation can also evolve.

**Plumage** No two barn owls are alike. Some are dark reddish, some are white, and others are speckled with many large black feather spots. This variation is observed between and within populations and even between siblings. Melanin is the most common pigment in animals, and it has multiple functions. It confers camouflage, protects the skin and feathers from biophysical degradation and ultraviolet light, participates in thermoregulation, and signals aspects of quality to potential mates and conspecifics. The study of melanin-based colouration has become a major focus in evolutionary ecology, and to this end, the barn owl is a prime model system. In this species, differently coloured individuals adopt alternative reproductive strategies and behaviours.

**Population dynamics** Annual variations in population size are pronounced in the barn owl. Studying the processes underlying variation in population size requires species that are particularly sensitive to environmental factors, such as inclement weather and variation in food supply. Compared to those of other raptors, such as the kestrel, which exploits similar reproductive and foraging habitats, barn owl population sizes vary to a much higher degree. This difference raises a number of questions about why this species is one of the first to suffer under ecological deterioration and how individuals compensate once conditions improve by reproducing at a high rate. Because the barn owl is cosmopolitan and often lives close to humans, this species is suitable for monitoring the impact of human activities on wildlife and habitat degradation worldwide.

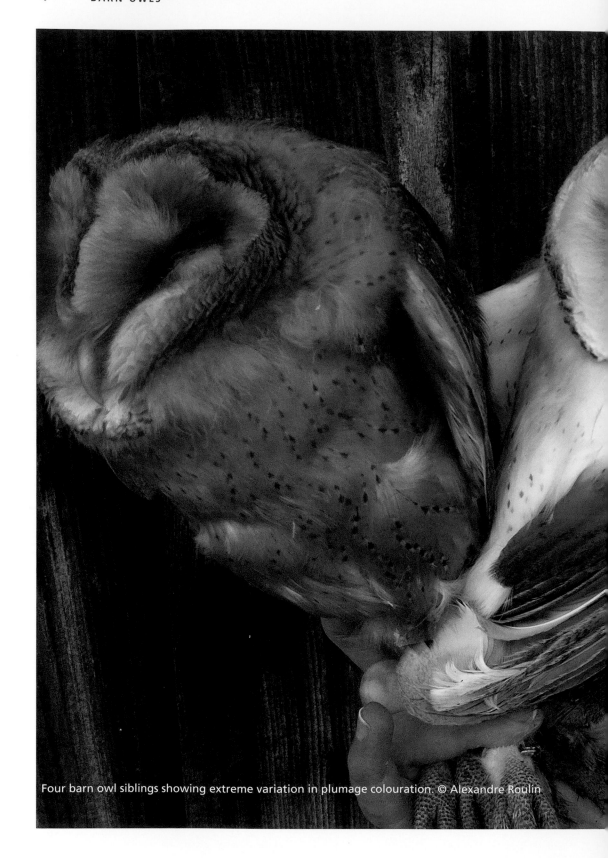

Four barn owl siblings showing extreme variation in plumage colouration. © Alexandre Roulin

The mouse is a model system for laboratory medical studies.

## 1.2    WHY STUDY BARN OWLS INSTEAD OF LABORATORY MICE?

# An emerging model system

**The barn owl is beautiful, emblematic and scientifically interesting. Bird lovers agree that this bird should be studied in depth and protected, but why should funding agencies finance studies using the barn owl rather than biomedical or molecular studies using laboratory mice?**

In science, fields such as genetics, immunology and cell biology make significant advances using mice, flies and rock cress. These model organisms are so well studied and easy to work with that the study of any other organisms can appear counterproductive to high-profile researchers. Science, biomedicine in particular, invests most of its human and financial resources in research based on model organisms; thus research on other species requires justification. Although model organisms have indeed provided invaluable scientific insights, their study also results in substantial limitations to obtaining a general understanding of nature.

Organisms evolve through natural selection, which ultimately modifies the frequency of genes present in the ancestral population. It is therefore necessary to consider the evolutionary history of each species to understand why a given adaptation initially evolved. For instance, owls living on an island may display a whitish plumage not because it confers camouflage but because the owls that initially colonized this island happened to be light rather than dark. Evolutionary ecologists, however, are mainly interested in understanding whether physiological, morphological and behavioural adaptations have evolved to address specific ecological factors. Therefore, evolutionary ecologists prefer to study organisms in their natural environment rather than in the laboratory.

My aim here is not to discredit laboratory studies in animal models but to explain why studying another species, such as the barn owl, in its natural habitat is complementary. To this end, I contrast laboratory and field studies using four arguments.

**Artefact** Animals living outside their natural environment may behave in strange ways. Furthermore, the adaptive function of specific traits may not be obvious when they are studied under laboratory conditions. Only studies of animals in their natural environment, such as studies on wild barn owls, can ultimately reveal the context under which adaptations have evolved.

## Laboratory animals are not representative of wild counterparts
Laboratory animals cope well with the stress induced by laboratory conditions, have a short generation time, and are prolific. Because all laboratory animals share these properties, scientific results based on studies using these animals may not necessarily be of general applicability, but rather are restricted to animals that are living under stressful laboratory conditions.

**Artificial selection** Animals bred for many generations evolve new behavioural, physiological and morphological traits adapted to laboratory rather than natural conditions. Conversely, barn owls have evolved in the wild, interacting with their biophysical environment and other wild animals. Studies of this bird can therefore provide important ecological insights with general applicability.

**Biodiversity** Nature finds many ways to solve a single problem. Science will repeatedly rediscover the same set of solutions if using the same set of organisms. Working with organisms other than those usually studied by most researchers will be refreshing, and will offer novel insight into old problems. Deciphering the complexity of ecosystems, and understanding life in its full diversity, therefore necessitates studying a variety of species in addition to those used in the laboratory.

Thousands of papers and numerous books have been written on the barn owl. Library at the Swiss Ornithological Institute in Sempach. © Alexandre Roulin

## 1.3  THE RAW DATA

# The barn owl gets a lot of attention!

This book is based on scientific literature published between 1853 and 2018 on the barn owl and its relatives. I searched for all papers published in international and local journals. This necessitated visiting libraries, contacting authors directly and searching the web. I explicitly searched for papers about the ecological and evolutionary aspects of barn owls, with no geographical restriction.

While I researched 3696 papers and books on the Tytonidae family (many of them not available in the *Web of Science*), there are still at least 650 published papers left to be referenced! The graph opposite illustrates the number of papers published about Tytonidae studied in different parts of the world. Approximately 61% of studies of barn owls have focused on the European population of western barn owls, with most studies stemming from fieldwork conducted in Germany (564 papers, 15.3%), followed by studies of the American barn owl (492 papers, 13.3%). Therefore, the greater availability of information on European or American barn owls is likely the result of a greater research effort in Europe and the USA than in Africa, the Middle East, Asia, Oceania, the Pacific islands, Central America and South America. My hope is that this book will stimulate further research worldwide.

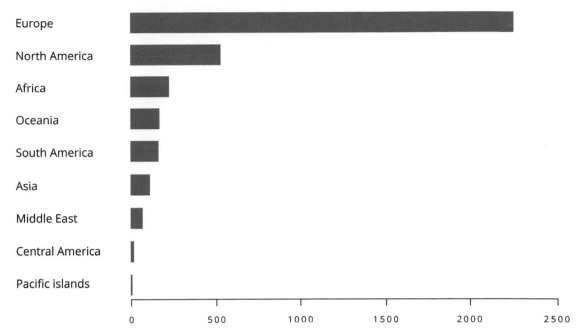

Number of scientific papers on Tytonidae species from different areas of the world.

Some *Tyto* taxa are more studied than others, implying that the information reported in the present book is biased towards these taxa. For this reason, this book mainly draws on research carried out on the common barn owl (traditionally referred to as *Tyto alba*) unless otherwise stated. The term Tytonidae refers to all species of the genera *Tyto* and *Phodilus*.

A barn owl in an old temple in India. An iconic view of how close the barn owl is to humans.

## 1.4   EVOLUTION OF THE TYTONIDAE

# The world is yours

**The family Tytonidae most likely first appeared in Australasia. This family has successfully colonized all continents except Antarctica. Australasia contains the most diverse populations of Tytonidae.**

Owls are the sister group to a diverse land-bird clade including mousebirds, trogons, cuckoo roller, coraciiforms and piciforms. Interestingly, all these land birds evolved from predators. Tytonidae diverged from typical owls (Strigidae) approximately 45 million years ago. At least thirteen Tytonidae taxa are extinct, and some were even larger than the imposing Eurasian eagle owl. For instance, the humerus of the extinct *Tyto gigantea* found in Italy measured 185 mm, compared to 73–85 mm in the contemporary Italian barn owl (*Tyto alba*).

**Tytonidae evolution and world colonization**  The available genetic data indicate that the ancestors shared by Strigidae and Tytonidae probably existed around the middle Eocene (about

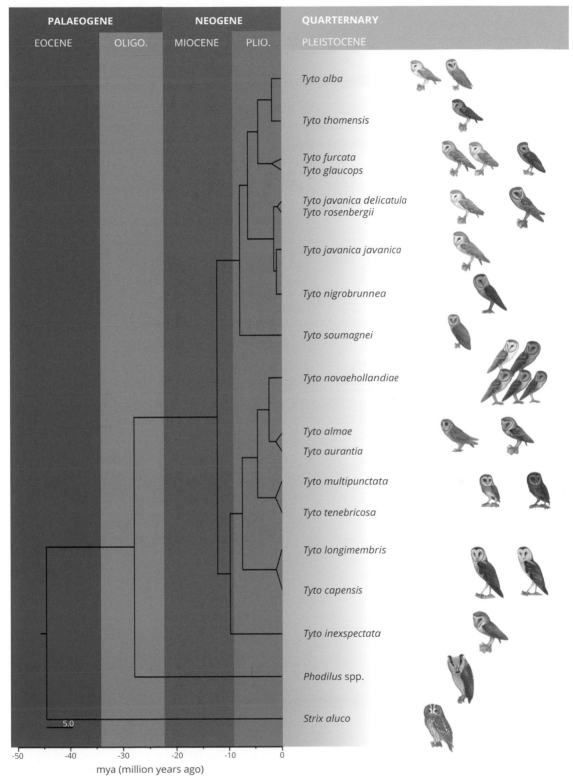

Phylogenetic tree of the Tytonidae family, with the tawny owl (*Strix aluco*) as an outgroup. The drawings are from the *Handbook of the Birds of the World Alive* (ed. J. del Hoyo *et al*., 2017). Lynx Edicions, Barcelona. (Retrieved from www.hbw.com in August 2018).

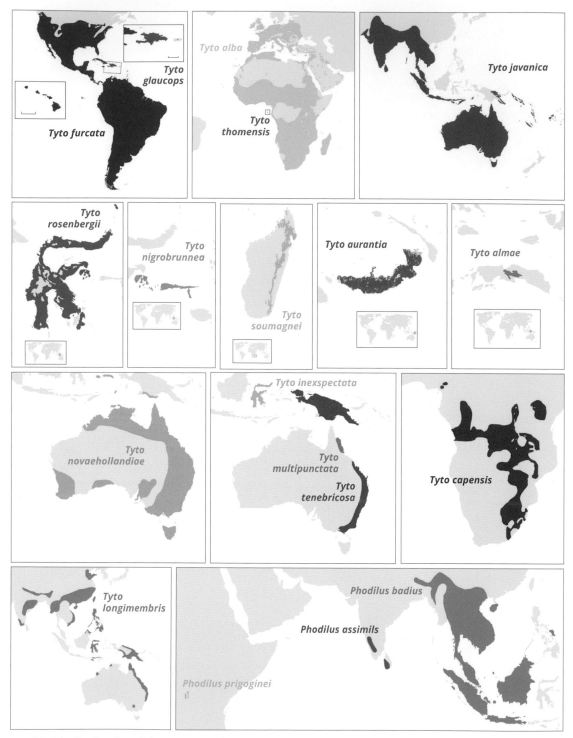

Worldwide distribution of the Tytonidae family. This includes American barn owls (*T. furcata* and *T. glaucops*), western barn owls (*T. alba* and *T. thomensis*), eastern barn owls (*T. javanica*), red owl (*T. soumagnei*), masked owls (*T. aurantia*, *T. almae*, *T. novaehollandiae*, *T. inexspectata*, *T. rosenbergii* and *T. nigrobrunnea*), sooty owls (*T. tenebricosa* and *T. multipunctata*), African and eastern grass owls (*T. capensis* and *T. longimembris*) and African and Australasian bay owls (*Phodilus prigoginei*, *P. assimils* and *P. badius*).

45 million years ago, mya), and that the most recent common ancestor between the two extant Tytonidae genera, the bay owls (*Phodilus* spp.) and the barn owls, grass owls and masked owls (*Tyto* spp.), was from the Oligocene (about 28 mya). The ancestor of the core *Tyto* (masked, sooty and grass owls) and the common barn owl group was from the Middle Miocene, about 12 mya, and first occurred probably in Australasia. From there, about 8 mya, an extinct ancestor dispersed and gave birth to the common barn owl group, currently known as the eastern barn owl (*Tyto javanica*), western barn owl (*T. alba*) and American barn owl (*T. furcata*), which reached the Americas via the North Pacific Bering land bridge.

Within each of these three lineages, some populations are isolated on remote islands, with little contact with other populations through dispersal. This facilitates genetic differentiation leading to the evolution of new species. For example, the Sulawesi masked owl (*Tyto rosenbergii*) and Taliabu masked owl (*T. nigrobrunnea*) are located on islands and are considered, by some authors, different species from Australasian barn owls (*T. javanica*). Similarly, the ashy-faced owl (*T. glaucops*) located on Hispaniola is a species on its own rather than a subspecies of the American barn owl (*T. furcata*) as previously considered.

## Europe

Traditionally, ornithologists recognize four European subspecies of the western barn owl (*Tyto alba*): the 'white-breasted barn owl' (*T. a. alba*) in southern Europe and on the British Isles, the 'dark-breasted barn owl' (*T. a. guttata*) in northeastern Europe, the 'Sardinian barn owl' (*T. a. ernesti*) in Sardinia and Corsica and the 'Middle Eastern barn owl' (*T. a. erlangeri*) in Crete and Cyprus. In Europe, genetic differentiation between countries is limited, indicating pronounced genetic exchange between populations through dispersal.

During the last Ice Age, barn owls, along with numerous other animals and plants, disappeared from most of the continent. Once the ice retreated, barn owls recolonized Europe through two routes in a ring-like manner. From the Iberian Peninsula they reached central and northeastern Europe, and from the Jordan valley they reached Greece and moved towards northeastern Europe. This scenario is based on genetic data showing that owls from the Jordan valley are genetically closer to owls found in the Canary Islands and Spain than to geographically closer populations from eastern Europe and the Balkans. Birds from these two lines currently hybridize in the Balkans, and although they originate from the same place in the Jordan valley (or possibly somewhere in North Africa), they may show genetic incompatibility. In other words, when birds from these two lines hybridize in the Balkans, they may produce offspring that have some genetic problems preventing them from growing optimally or surviving for long. This is an interesting hypothesis to tackle in the future.

## Africa

In sub-Saharan Africa, there are two species of *Tyto*, the western barn owl (*T. alba*) and the African grass owl (*T. capensis*), the latter being bigger and with longer legs. The biology of the barn owl in Africa is similar to that of barn owls in other parts of the world; it breeds in cavities and produces one or two annual broods. In contrast, the African grass owl breeds on the ground in tall grasses. The Madagascan red owl (*T. soumagnei*) shares an extinct ancestor with the western barn owl, which was located between Africa and Australasia. We now recognize that the barn owls located on São Tomé Island (*T. thomensis*) have genetically diverged sufficiently from their continental counterparts (*T. alba*) to be potentially considered a different species. In contrast, barn owls from Cape Verde (*T. alba detorta*) and the Canary Islands (*T. a. gracilirostris*) display low genetic divergence, and are still considered subspecies of *Tyto alba*.

## Asia

In Asia, ornithologists recognize two groups of *Tyto*, the eastern barn owls (*T. javanica*) and the grass/masked owls. The white-coloured Middle Eastern population of the western barn owl (*T. alba erlangeri*) is found in the Arabian peninsula, the Jordan valley, Iran and Iraq and is genetically closer to the western barn owl than to the darker and spottier Indian population of eastern barn owl (*T. javanica stertens*) found in Pakistan, India and Sri Lanka. In Southeast Asia (Thailand, Vietnam), the closely related eastern barn owl is found: it is larger and has very diverse plumage patterns. Although the Sulawesi masked owl (*T. rosenbergii*) is quite large, it is closely related to eastern barn owls rather

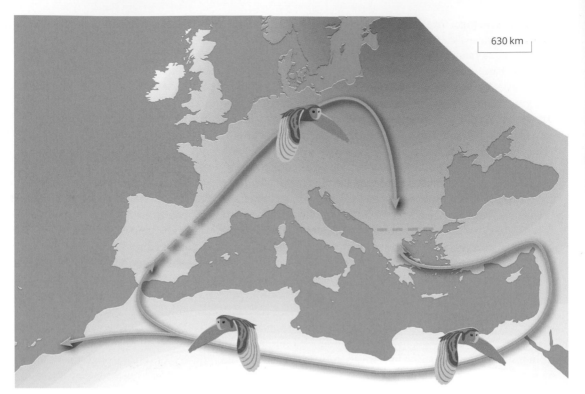

Circum-Mediterranean ring-like colonization of Europe by barn owls. After the last Ice Age, barn owls retreated to the Iberian Peninsula, North Africa and the Middle East. After the ice retreated 10 000 years ago, barn owls recolonized Europe through two fronts: one from the Jordan valley and another from the Iberian Peninsula. The zone of contact between owls from these two colonization routes is in the Balkans. The colour gradation indicates the extent to which the underparts of European barn owls are reddish.

than to masked owls. In fact, because the Malay Archipelago between mainland Southeast Asia and Australia contains more than 25 000 islands, animals quickly diverge into different forms on different islands that are geographically distant from each other, at least sufficiently to limit frequent exchanges of individuals. As a consequence, new forms or species are regularly described on remote islands. This extreme level of diversity makes Asia particularly interesting for the study of barn owls.

In the second group, there are the grass owls and masked owls. The eastern grass owl (*Tyto longimembris*) is much larger and has longer legs than barn owls. This species is found from India to China and on many islands leading to Australia, New Caledonia and Fiji (although it recently disappeared from Fiji). Despite having a different scientific name and being geographically distant, *T. longimembris* is genetically and biologically strikingly similar to the African grass owl (*T. capensis*). This is a fine example of a rapid radiation between distant continents that were once geographically close. A number of masked owls – such as the Seram masked owl (*T. almae*), golden masked owl (*T. aurantia*), Minahassa masked owl (*T. inexspectata*), Taliabu masked owl (*T. nigrobrunnea*) and Moluccan masked owl (*T. sororcula*) – are also found in the Malay Archipelago. Their phylogenetic relationships still require attention, but we already know that the Taliabu masked owl is genetically closer to the eastern barn owl than to other masked owls.

**Oceania**  Like the Malay Archipelago, Oceania is a hotspot of diversity. In Australia there are two distinct groups of *Tyto*, as in Asia. From Australia, the eastern barn owl (*Tyto javanica delicatula*) has spread to numerous Pacific islands, reaching American Samoa and Niue, and is currently colonizing New Zealand. Although the Asian population of eastern barn owl (*T. j. javanica*) belongs to the same species as the barn owls from Oceania, the two forms have slightly diverged genetically.

The eastern grass owl (*Tyto longimembris*) is found in Australia and New Guinea, while the Australian masked owl (*T. novaehollandiae*) is also found in Tasmania (*T. n. castanops*). Sooty owls (*T. tenebricosa*) uniquely display completely black plumage and exploit forests in eastern Australia (*T. t. tenebricosa* and *T. t. multipunctata*) and New Guinea (*T. t. arfaki*). Some authorities treat the sooty owl as two species (*T. tenebricosa* and *T. multipunctata*), but further work is required to clarify the exact phylogenetic relationships among sooty owls. More generally, Oceania deserves attention to explain such vast diversity in an otherwise quite homogeneous family.

**Americas**  In North, Central and South America, the only recognized *Tyto* species are the American barn owl (*Tyto furcata*) and the ashy-faced owl (*T. glaucops*) from Hispaniola. The exact phylogenetic relationships between different populations and the extent to which they have genetically diverged are not yet known. The Caribbean region is particularly interesting, as barn owl morphology has significantly diverged between nearby islands. On the northern American continent, the different populations are genetically quite homogeneous, much more than in Europe. This indicates that there are many more exchanges of individuals between populations in North America than in Europe, which is consistent with the fact that American barn owls disperse more than their European counterparts and even migrate, which is apparently not the case in Europe. However, the Rocky Mountains act as a barrier to dispersal between owls from the west coast and those from the Midwest and the east coast.

## FUTURE RESEARCH
- The recent development of genomic tools permitting the screening of thousands of genes will offer new insights into the phylogenetic relationships among Tytonidae, and will also help to identify the routes along which barn owls spread worldwide.
- After the last Ice Age western barn owls recolonized Europe from two fronts, one in the Middle East towards Greece and the other from the Iberian Peninsula in the direction of Germany. These two lines of barn owls were initially in contact before the last Ice Age, separated during the recolonization process, and are now again in contact in the Balkans – a so-called secondary contact. The separation may have been sufficiently long for the two lines to diverge genetically. Genetic analyses should be performed to test whether there are genetic incompatibilities in the hybrid zone between owls that originated from the Iberian line and those from the Middle Eastern line.

## FURTHER READING
Antoniazza, S., Ricardo, K., Neuenschwander, S., Burri, R., Gaigher, A., Roulin, A. and Goudet, J. 2014. Natural selection in a post-glacial range expansion: the case of the colour cline in the European barn owl. *Mol. Ecol.* **23**: 5508–5523.

Burri, R., Antoniazza, S., Gaigher, A., Ducrest, A.-L., Simon, C., The European Barn Owl Network, Fumagalli, L., Goudet, J. and Roulin, A. 2016. The genetic basis of color-related local adaptation in a ring-like colonization around the Mediterranean. *Evolution* **70**: 140–153.

Machado, A. P., Clément, L., Uva, V., Goudet, J. and Roulin, A. 2018. The Rocky Mountains as a dispersal barrier between barn owl (*Tyto alba*) populations in North America. *J. Biogeogr.* **45**: 1288–1300.

Prum, R., Berv, J. S., Dornburg, A., Field, D. J., Townsend, J. P., Lemmon, E. M. and Lemmon, A. R. 2015. A comprehensive phylogeny of birds (*Aves*) using targeted next-generation DNA sequencing. *Nature* **526**: 569–573.

Taberlet, P., Fumagalli, L., Wust-Saucy, A.-G. and Cosson, J.-F. 1998. Comparative phylogeography and postglacial colonization routes in Europe. *Mol. Ecol.* **74**: 453–464.

Uva, V., Päckert, M., Cibois, A., Fumagalli, L. and Roulin, A. 2018. Comprehensive molecular phylogeny of barn owls and relatives (Family: Tytonidae), and their six major Pleistocene radiations. *Mol. Phylogenet. Evol.* **125**: 127–137.

Wink, M., Heidrich, P., Sauer-Gurth, H., Elsayed, A. A. and Gonzalez, J. 2008. Molecular phylogeny and systematics of owls (Strigiformes). In König, C. and Weick, F. (eds) *Owls of the World*, second edition. London: Christopher Helm, pp. 42–63.

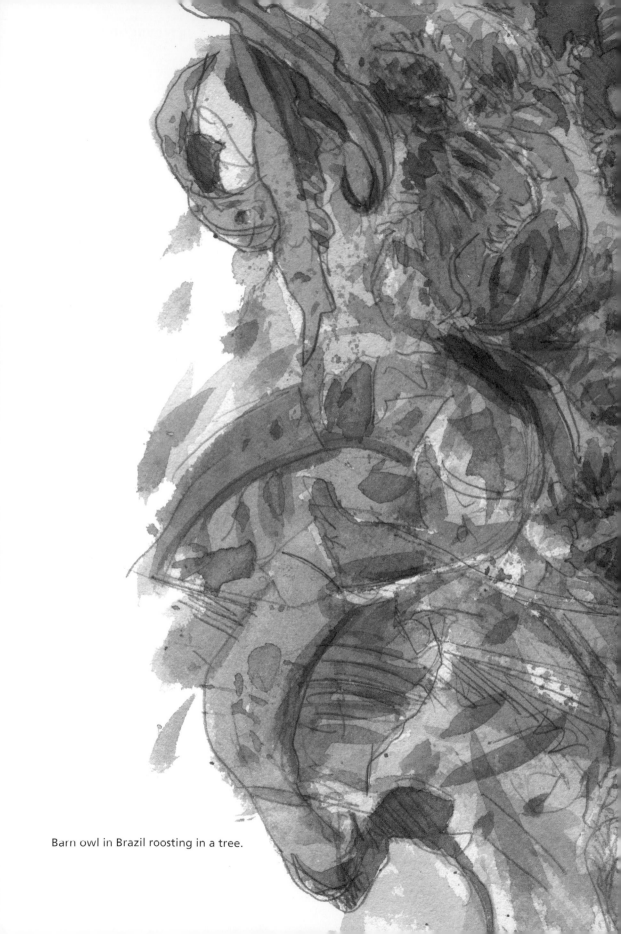

Barn owl in Brazil roosting in a tree.

Barn owl roosting in a palm tree in the Middle East.

## 1.5   WHY COSMOPOLITAN?

# A success story

The common barn owl (*Tyto alba*, *T. furcata* and *T. javanica*) is one of the few extant cosmopolitan animals. Successfully spreading throughout the world requires tolerance of a broad range of environments, competitiveness to occupy ecological niches, and the ability to nest in various locations and disperse effectively.

Only a few land animals are cosmopolitan, including the peregrine falcon, osprey, cattle egret, barn owl and humans. The exceptional cognitive capacity of humans has allowed us to explore almost all habitats and to survive under any conditions. Unfortunately, due to a lack of data, we are left to hypothesize as to why the barn owl has been so successful. I therefore discuss potential avenues of research to explain the cosmopolitan distribution of this animal. A number of adaptations are necessary for a nocturnal predator to spread throughout the world. What is so unique about the barn owl compared to other birds and, in particular, other nocturnal animals?

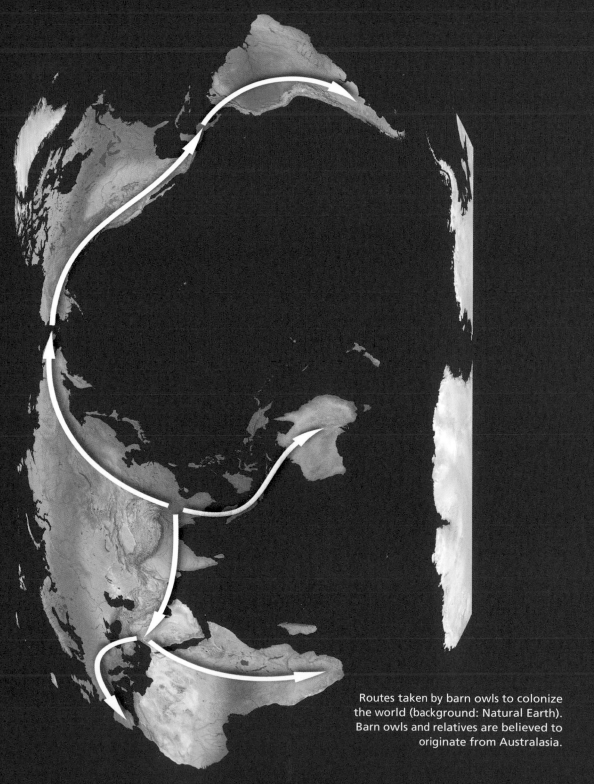

Routes taken by barn owls to colonize the world (background: Natural Earth). Barn owls and relatives are believed to originate from Australasia.

A flying barn owl in Madagascar.

**White plumage** In contrast to the snowy owl, which effectively blends into its white environment, the barn owl has not evolved white plumage for camouflage. Rather, the white colouration may have evolved to disturb prey that, in reaction to the sudden appearance of a conspicuous predator, does not adopt an efficient escape response. White plumage flashes into the visual field of a small mammal, which freezes for long enough to be caught before the flight response kicks in, an effect that is exacerbated under the full moon. Because small mammals rarely encounter a white predator, they do not have the opportunity to learn how to react. Most nocturnal predators rely on crypsis to hunt, and thus natural selection on small mammals is not strong enough to promote the evolution of an appropriate response to white predators, such as the barn owl. I therefore raise the hypothesis that white plumage has allowed the barn owl to forage efficiently worldwide. In situations where the advantages bestowed by white plumage disappeared, *Tyto* evolved different plumage patterns, reddish or even black, as in sooty owls.

**Hearing and not being heard** The barn owl flies so silently that mice hardly notice its presence, and the owl's super-efficient hearing and facial mask allow it to detect prey with surgical precision. The ability to fly silently is probably key to hunting while gliding, an ability shared with only a few other birds, such as harriers. Thus, wherever the barn owl lands, it can find prey, usually small mammals, which are found almost everywhere.

**Opportunistic diet** Not specializing in a single type of prey, the barn owl has a high dietary flexibility and hence resilience to changing prey availability. Encountering new types of prey was therefore probably not a problem for barn owls exploring new areas. Its moderate size also helped to fill this broad ecological niche.

**Supersonic evolution** Few barn owls disperse long distances, but enough to colonize new areas. After crossing a large body of water to reach new land, barn owls quickly evolved new morphological, physiological and behavioural adaptations suited to the newly discovered regions. This process is facilitated by the barn owl's tendency to remain in its area of birth. As a consequence, newly established populations could rapidly evolve new adaptations because the flux of immigrants with non-adapted behaviour and morphology was very low.

**Rapid differentiation but limited speciation** Despite an ability to rapidly adapt to new environments, American, Eurasian, African and Australian barn owls, masked owls and grass owls share a similar morphology and plumage. Therefore, the Tytonidae family did not diverge into very different species, suggesting that its body plan is close to perfection and should not be modified, at least not to a large degree. Thus, the ability of the barn owl and its relatives to colonize new environments was not based on the evolution of fundamentally new adaptations upon reaching new areas. Rather, barn owls had at their disposal the necessary pre-adaptations to colonize the world.

**Baby boom** For a bird this size, the barn owl is exceptionally prolific: it starts breeding in the first year of life and can produce up to three annual broods, each with up to ten fledglings. This makes the barn owl an r-strategist, breeding at a rate comparable to the much smaller passerines. This prolific reproduction certainly helped the barn owl fill empty ecological niches in newly colonized areas.

The ability of the barn owl to exploit anthropogenic environments can certainly explain the expansion of its range within continents, but not its worldwide spread. The barn owl expansion occurred long before agriculture first appeared 10 000 years ago. From this time onward, humans cleared forests, opening new open habitats in which the barn owl could reproduce and forage. Human activities therefore provided new opportunities for the barn owl to spread in areas that were previously inhospitable, such as in Ohio in the USA where not enough open habitats were available. Agriculture has also increased the availability of food for rodents and, in turn, their predators, including the barn owl.

Everything has a limit, however, and traits favouring one currency often occur at the expense of another currency. While the barn owl is indeed a successful colonizer, its distribution does not extend to the northernmost reaches of North America and Eurasia, or to Antarctica. And although the barn owl is very sensitive to cold climes, it rarely migrates to more clement regions in winter. I speculate that these two characteristics (sensitivity to harsh winters and not migrating) played fundamental roles in shaping the barn owl's distribution, the other side being all other characteristics that allowed this owl to colonize the world. Being resistant to cold weather may require morphological, physiological and behavioural adaptations that are not compatible with the ability to spread throughout the world.

## FURTHER READING
San Jose, L. M., Séchaud, R., Schalcher, K., Judes, C., Questiaux, A., Oliveira-Xavier, A., Gémard, C., Almasi, B., Béziers, P., Kelber, A., Amar, A. and Roulin, A. 2019. Differential fitness effects of moonlight on plumage colour morphs in barn owls. *Nature Ecology & Evolution* **3**: 1331-1340.

Barn owl entering a farm building with prey in its bill.

## 1.6   WHY DO BARN OWLS LIVE SO CLOSE TO HUMANS?

# Synanthrope

**Nature has suffered from human domination. Over thousands of years, humans have hunted many wild animals or relegated them to remote places. In certain cases, however, anthropogenic activities have created plentiful sources of food for certain species and removed predators of a few non-domesticated animals, such as crows, gulls and rats. While some barn owl populations have benefited from human activity, disturbance by people is becoming a serious problem for most Tytonidae, one that will become greater in time to come.**

The relationship between humans and animals has a very long history and can be traced back to the time of the Pharaohs, the ancient Greeks and the Chinese dynasties. Human buildings offered new breeding sites to a number of birds including the house sparrow, house martin and barn owl. Humans have developed a special love–hate relationship with the last of these. Barn owls have fascinated us since the dawn of humanity, for various reasons – including their tendency to breed in human constructions, their nocturnal habits, their scary calls, their angelic or ghost-like white plumage and their flat faces.

The barn owl originally mainly bred in cavities within cliffs and trees. However, human constructions have acted as suitable alternatives, allowing the range of this species to spread alongside humans.

Barn owls are found even in large cities.

Thus, barn owls are one of the few animals that have benefited from human activities, in contrast to other Tytonidae that breed on the ground (grass owls), in forests (sooty owls and a few others) or in large tree cavities (masked owls). Breeding so close to humans also necessitated the ability to cope with the stress generated by humans and their daily routines.

When early humans switched from hunting and gathering to agriculture, the production of wheat and other seeds favoured small mammals, the main food source for barn owls. Humans cleared forests, which proved favourable to the barn owl, a bird that forages in open landscapes, and buildings offered new nest sites and roosts. However, the situation is changing dramatically. As old buildings are replaced or renovated, they offer few possibilities to breed, and intensive agriculture is reducing the diversity of available crop food sources for small mammals, as well as being associated with copious inputs of toxic rodenticides, which further impact the barn owl's prey and poison the owls. Furthermore, high road traffic density increases the risk of collision for barn owls, which are particularly susceptible given their tendency to fly low over the ground.

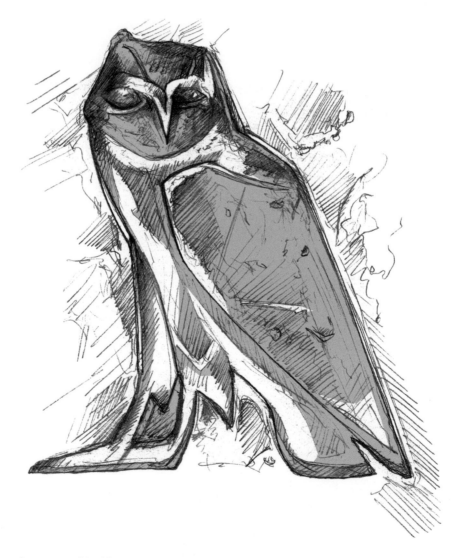

The barn owl was a mythical beast in ancient Egypt.

# 2 Conservation

2.1 Why protect barn owls?
2.2 Ethics
2.3 Decreases in barn owl populations
2.4 Pollution
2.5 What can we do to protect barn owls?

Forests are destroyed worldwide, as here in Tasmania. © Michael Todd.

## 2.1   WHY PROTECT BARN OWLS?

# It's never too late to repair the irreparable

**Although ornithologists are motivated by an urgent need to protect nature, we need to convince society as a whole. Why invest money and human resources in nature conservation? Why preserve barn owls?**

The anthropocentrism philosophy considers the rise of humans as the dominant species on earth to be justifiable. Three arguments are presented: (1) Human supremacy over all other organisms is the wish of a superior force (God), and we are the vehicle of this divine intervention; (2) Our intellectual superiority gives us the right to dominate all other inferior organisms; and (3) Humans have faced natural selection over thousands of years, as have other organisms; thus, humans rule the world as dinosaurs did millions of years ago, and so what? The negative impact of humanity on the environment may therefore appear 'natural' or legitimate, as long as it is useful to mankind. The paradox, however, is that what is useful today might be disastrous tomorrow.

While deforestation provides wood and space to cultivate fields, excessive tree-cover loss has severe impacts on ecosystem services including the depletion of water resources, decreased resorption of $CO_2$ and soil degradation, not to mention the loss of biodiversity. Another example is the use of DDT: this chemical kills insects to increase plant production but eradicates the beneficial insects that pollinate these plants. A scientific solution would be to produce genetically modified pollinators that resist insecticides; but this approach may have unintended negative impacts. Science is capable of developing short-term, often complex and clever solutions, but scientists often do not appreciate

Food chain: a barn owl holding a wood mouse in its bill, while the rodent has an ear of corn. Photo taken with a camera trap in Switzerland. © Robin Séchaud & Kim Schalcher

or investigate the full societal and ecological ramifications of these solutions, and the long-term impacts may even be impossible to predict.

An economic model that engenders greed and the over-exploitation of natural resources, favouring economic growth, can indeed generate gains in the short term, but we can predict the depletion of natural resources and the extinction of biodiversity stemming from our current economic obsession. Before the multitudes of species in the wild have disappeared, with only a few individuals left roaming in dedicated sanctuaries and zoos, the point of no return will have been reached. Biodiversity provides humans with high-quality resources and the healthy environment we depend upon. Successful coexistence with other species will be a testament to our ability to live sustainably, within ecological bounds. It is therefore our duty to protect nature.

## Ecological function of the urban perspective
Most humans live in cities, disconnected from both the natural environment in which our species initially evolved, and the habitats society depends on. Most people do not even consider how water arrives at their taps, as if easy access to water is the normal state of affairs. Most humans wonder 'why should we protect nature?' – and if they gave it a moment's thought they might also ask 'why in particular should we protect the barn owl?' The answer is that top predators such as the barn owl stabilize ecosystems by structuring lower trophic levels. In North America, the reintroduction of wolves has balanced deer populations and, in turn, forest growth and composition. In Europe, wild boars, which cause problems for agriculture, thrive because their natural predators, such as wolves, are killed. Biodiversity can also be preserved simply to meet our need for knowledge, but conservation is synonymous with our ability to live within limits – and ultimately with our survival.

For biologists seeking to implement effective conservation measures, the acquisition of detailed knowledge is the necessary first step. However, when faced with applications for the use of public money for conservation research, urbanites are often unsupportive and further arguments are needed. Watching a good TV programme on animals may be sufficient to engender the feeling that 'we love nature' provided we do not have to participate directly or indirectly in nature conservation. For some, nature may have an aesthetic or ludic value: forests are used for relaxing, and a cliff is merely a place to perform extreme sports, such as climbing or BASE jumping. Of course, nature is much more than this and does not belong to humans simply for recreational, or any other, purposes.

## Owls for peace

At least ten million species live on earth. Thus, why should we care about the barn owl specifically?

This bird fascinates us, stimulates our imagination, yet scares us and is tied to myths which have existed for millennia. Stories of the barn owl's natural history generate empathy for the catastrophic impact of global changes caused by human activities. For countless years, nature has inspired human development in science, fashion, the arts and many other fields. The barn owl is no exception, having led to developments in the fields of aeronautics, neuroscience and biomedicine. This species is also an ideal subject for education, particularly to increase awareness of the ecological importance of protecting nature. The bones of the barn owl's prey are exceptionally well preserved in pellets, providing a unique opportunity to study predator–prey interactions through a fun pellet dissection!

The barn owl can even act as a gateway to peace. Barn owls have brought members of the Israeli, Palestinian and Jordanian communities to the same table. In this zone of intense conflict, people-to-people interactions are necessary to maintain the dialogue between war-torn communities. In the Middle East, as in many other regions, farmers spread substantial quantities of rodenticides to kill the voles and mice that devastate their crops. After numerous discussions, meetings and conferences, farmers have chosen to favour barn owls as biological pest control agents instead of using copious amounts of rodenticides. This represents a cultural shift, and a huge success, given that many farmers previously believed the barn owl to be a bad omen. What is bringing Israeli, Palestinian and Jordanian farmers to sit around the same table is a shared common problem, namely the issue of

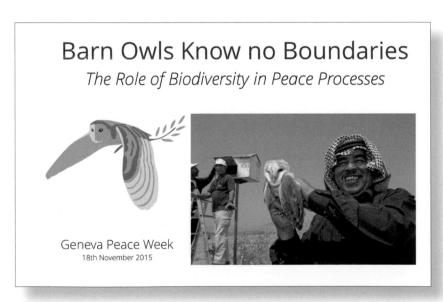

The barn owl is a flagship species for nature conservation. A project using barn owls as biological pest control agents in the Middle East was presented at 'Geneva Peace Week' on 18 November 2015 at the House of Peace at the Geneva Centre for Security Policy. This conference emphasized the role played by nature conservation in the reconciliation between Israelis, Palestinians and Jordanians.

agricultural pests, and the need to find a solution that is both economical and ecological. This is an issue that 'knows no boundaries', given that mice and owls certainly do not respect artificial human political borders. This approach successfully brings people together because it avoids the sensitive topics, such as culture, tradition and religion, that are at the root of the conflict, and instead provides common ground for dialogue.

In summary, the barn owl has all the qualities to be considered a flagship species for nature conservation as a whole, and may even become a symbol of peace. In this sense, the human–barn owl relationship embodies what Bobby McLeod, an Aboriginal activist and poet said – '*To heal ourselves we must heal our planet, and to heal our planet we must heal ourselves.*'

## FURTHER READING

Roulin, A., Mansour, A. R., Spiegel, B., Dreiss, A. N. and Leshem, Y. 2017. 'Nature knows no boundaries': the role of nature conservation in peacebuilding. *Trends Ecol. Evol.* **32**: 305–310.

Sergio, F., Schmitz, O. J., Krebs, C. J., Holt, R. D., Heithaus, M. R., Wirsing, A. J., Ripple, A. J., Ritchie, E., Ainley, D., Oro, D., Jhala, Y., Hiraldo, F. and Korpimäki, E. 2014. Towards a cohesive, holistic view of top predation: a definition, synthesis and perspective. *Oikos* **123**: 1234–1243.

Smith, D. W., Peterson, R. O. and Houston, D. B. 2003. Yellowstone after wolves. *BioScience* **53**: 330–340.

Scientific studies often require the capture
and manipulation of wild barn owls.

## 2.2   ETHICS

# The end does not necessarily justify the means

**Implementing effective nature conservation measures requires that we observe animals in their natural habitat. However, the mere presence of ornithologists can sometimes represent an intrusion into the daily lives of animals. Often, we face the paradox that protecting what we love may require disturbing it. How do we resolve this?**

Working with animals bestows upon us the duty to adopt an ethical approach. This is a complex problem because we must reconcile the need to gather scientific data to promote nature conservation with the need to avoid disturbing animals. Certain people believe that as long as animal populations are not jeopardized by scientific observations, human interference can be tolerated even if it negatively affects some individuals. Others claim that the act of capturing one animal for scientific purposes is an intolerable intrusion.

Assuming that animals do not suffer in any measurable way because of scientific work, and assuming that reproductive success is not penalized by that work, it can be difficult to define what 'disturbance' means without introducing psychological terms. Most often researchers are careful to minimize their impact, to ensure that individuals suffer no long-term effects, and that their reproductive success is not penalized. However, given the difficulty in assessing disturbance, and even defining what we mean by disturbance, it is hard to understand the extent to which ecologists might be affecting individuals.

**Experimental approach** Science uses different approaches to answer questions about animal behaviour and ecology. The most controversial approach is the **experimental method**, in which animals are manipulated. Experimentation is not the only way biologists answer questions about the natural world, but it is a tool that is used by many ecologists. The logic of **experimental design** is to manipulate a factor in order to measure its effect independently of any other variables. With everything else being equal, the only thing that distinguishes individuals in a study is the experimental treatment.

For instance, to determine whether barn owl parents adjust the number of prey items brought per night per offspring in relation to brood size, we can experimentally manipulate brood size. To decouple habitat quality from the parental ability to modify feeding rate in relation to family size, we can exchange a different number of hatchlings between randomly chosen nests. For instance, two hatchlings from nest A are put in nest B and one hatchling from nest B is put in nest A. If we replicate this exchange of two nestlings with one nestling between many pairs of nests, we will have a representative sample of nests for which brood size has been experimentally enlarged or reduced by one nestling. Even if parents initially intended to raise a given number of nestlings, we force them to take care of one extra or one fewer chick. This experimental design is feasible in birds because parents are usually unable to identify their progeny. A couple of weeks later, we can measure parental feeding rate, with the prediction that parents tending an experimentally reduced brood will bring more food items per night per nestling than will parents tending an experimentally enlarged brood.

Why is it not enough to measure parental feeding rates in relation to brood size in non-manipulated nests? Simply because it is difficult to propose *a priori* predictions in a non-experimental set-up. Each nestling may be fed at a lower rate in larger broods because food resources are limited and must be shared among more mouths. Alternatively, each nestling may receive more food items in larger

Negotiation calls are broadcast to two barn owl siblings in a so-called 'playback experiment'.

broods if stronger parents or parents living in better habitats are able to produce larger broods. These parents could raise more offspring and simultaneously feed all of them at a higher rate than could weaker parents living in poor environments.

To the lay public, the word 'experiment' is often associated with cruel treatments inflicted on animals. However, for the evolutionary ecologist seeking to understand why certain physiological, morphological or behavioural traits evolved in a particular environment, experimental design must remain within the natural range of variation for the variable of interest. Thus, creating experimental broods containing sixteen chicks when the maximal brood size in the wild is twelve makes no sense. Observing that parents tending sixteen nestlings feed each chick at a lower rate than do those tending five-chick broods tells us nothing about whether parents increase hunting effort for their offspring at the expense of self-feeding. Experiments should always place the species in a realistic context that is no more 'cruel' than the situation that they experience daily.

Implementing appropriate experimental designs is not trivial. In the 'brood-size manipulation experiment' above, we must allocate different brood sizes randomly across parents and habitats, because it is likely that stronger parents produce more chicks. Thus, if we want to measure the effect of brood size independently of other variables, strong and weak parents across habitats of varying quality should be given the opportunity to raise a similar number of chicks.

Carefully designed experiments can answer an almost endless variety of questions that fascinate the human mind. However, in the face of the current 'sixth mass extinction event' and the second 'scientists' warning to humanity', perhaps it is time to concentrate our minds on seeking answers that will bring the greatest benefit for conservation and environmental education.

**Are scientific studies more useful to nature or to the scientists?** Evolutionary ecologists often work with wild organisms outdoors. In doing so, they join a variety of different stakeholders, such as farmers, hunters and nature lovers, who also utilize the outdoors in one way or another. The usefulness of observations can sometimes be questioned, and consequently scientists can be perceived to be using the natural environment for their own career interests, rather than to ultimately help preserve nature. This is why it is so critical that, as researchers, we communicate effectively and justify our work to the lay public.

Biodiversity is under intense pressure and many species are disappearing. This motivates some researchers and practitioners to invest effort and money in conservation using a variety of complementary approaches. The first method is to carry out applied research aimed at devising effective measures to preserve habitats and species. Another approach is to determine the factors that are limiting survival, productivity and dispersal. Fundamental research that is *a priori* not directly related to conservation can also be important because it just might, sooner or later, provide a missing bit of knowledge that can be translated into conservation action, assuming that researchers interact with practitioners.

The present book aims to summarise the current knowledge about the barn owl, a flagship species, to spread the noble message of conservation. An intimate knowledge of animals can help motivate society to take active measures to protect animals. '*In the end we will conserve only what we love; we will love only what we understand; and we will understand only what we are taught*' (Baba Dioum, 1968).

**Respectful methods** To successfully spread the message that we must protect nature, scientific credibility and the use of respectful methods while studying barn owls are essential. Experienced fieldworkers usually know what can and cannot be done while working with wild barn owls and how to minimize disturbance. Some essential practices need to be emphasized:

- Barn owls are particularly sensitive to disturbance during the pre-laying and laying period. Visiting nest sites before eggs are laid can induce owls to abandon sites, especially individuals that are new or underweight. To avoid this situation, ornithologists should prospect potential sites at night without capturing them.

Two Palestinian farmers holding a barn owl. © Hagai Aharon

- During incubation, owls can be sensitive to disturbance and abandon their clutch if captured, an effect that can differ between individuals and between populations.
- The ties between parents and offspring are stronger at the nestling stage than at the egg stage, and parents rarely abandon their young if disturbed. However, disturbance must still be kept to a minimum, as cases of human disturbance leading parents to stop caring for their brood have been observed, although primarily when the nestlings were very young.
- When parents are captured in the nest, care must be taken to ensure they do not fly away once they are placed back in it. An adult owl flying in daylight can be attacked by crows, and if the eggs and young are left unattended for a long period the nest can ultimately fail.
- Finally, the potentially negative impact of disturbing owls for the purpose of monitoring makes sense only if the research leads to implementation of effective conservation measures, and if ornithologists communicate their work.

In the nineteenth century, farmsteads in the United Kingdom could host dozens of barn owls. This is unfortunately no longer the case.

## 2.3   DECREASES IN BARN OWL POPULATIONS

# How many?

**Monitoring bird populations is one of the most important activities of ornithologists, particularly when species are becoming endangered. Some historical trends in barn owl population sizes are known in Europe and to a lesser extent in North America and Australasia. Unfortunately, these numbers indicate strong declines. There are only a few areas where populations are showing a substantial increase thanks to local conservation action.**

The arrival of European settlers in America in 1492 and in Australia in 1788 caused major ecosystem changes. In fact, in all 'developed' countries, nature has suffered profound habitat alteration and increased disturbance associated with human activities. Effects include increases in pollution, traffic, persecution, habitat degradation and intensification of agricultural practices.

Most cultures have a view on barn owls; in some they are considered a harbinger of evil or a source of black magic, in others they are used in traditional medicine, and in some they are even eaten! In 2008, a man was arrested in Malaysia for possessing 917 dead owls, of which 796 were barn owls, destined to be eaten in China. We certainly hope that such cases are one-offs and not indicative of an ongoing international trade. Ornithologists have worked hard to secure the legal protection of owls, something that was achieved only relatively recently worldwide (1954 in the United Kingdom, 1967 in Italy, 1971 in Australia, 1972 in the USA and Malaysia, 1973 in France).

Two pictures taken at the exact same place near Basel, Switzerland, in 1904 (© K. Lüdin) and 1999 (© Karl Martin Tanner). From Tanner, K. M. 1999. *Augen-Blicke. Bilder zum Landschaftswandel im Baselbiet. Quellen und Forschungen zur Geschichte und Landeskunde des Kantons Basel-Landschaft*, Band 68, Verlag des Kantons Basel-Landschaft, Liestal.

In more recent times, there can be little doubt that habitat loss is a much bigger factor than direct persecution. The situation is particularly problematic for habitat specialists, such as the sooty owls and grass owls, which suffer from forest clearance and intense use of grasslands for agriculture respectively. These species very rarely or never use nest boxes, so providing nest boxes is not an applicable solution to help restore their populations. This is anyway not the long-term answer for any of the Tytonidae – or any other birds.

The number of barn owls has been estimated in Europe, where between 100 000 and 250 000 pairs are reported, with most individuals being located on the Iberian Peninsula. Historical changes in the sizes of European barn owl populations illustrate the critical situation of this bird in many countries. In the Netherlands, for example, 1800–3500 pairs were breeding before 1963, 230–490 from 1969 to 1974 and 300–500 pairs from 1982 to 1984. In England and Wales, landscape modification caused a huge decrease in the barn owl population. Approximately 12 000 pairs were breeding in 1932, 4500–9000 from 1968 to 1972 and only 4000 in 1995 to 1997 across the entire United Kingdom.

In North America, data on barn owl population size are restricted to the northern range of the barn owl's distribution. For example, in Ohio, the first barn owl was recorded around 1860 when humans started to clear forests, offering new foraging habitats to this bird. Populations peaked in 1931–1935, and as crop fields progressively replaced less intensive forms of production, barn owl numbers decreased. Similarly, in British Columbia, Canada, the first barn owl was recorded in 1909;

the population reached 1000 pairs in 1983 and decreased to 250–500 pairs in 2008. Thus, owl populations can be both positively and negatively impacted by human activity.

Limited data are available for other continents. In Tasmania, the masked owl population decreased by a factor of two after human settlement. Because, like the grass owl, this owl very rarely breeds in nest boxes, preserving natural nest sites is even more crucial – as without them the owls are left with nowhere to breed. The situation in Malaysia is particularly interesting because it highlights the double-edged effect of human activities. Arriving from Java or Sumatra, barn owls started to breed in Malaysia at the end of the nineteenth century. Before 1968, there were few observations of breeding barn owls, but since the increase in rat numbers in newly established palm oil plantations, owls have been positively encouraged, through the provision of nest boxes, in order to control the rat population. Clearing rainforest for palm oil plantations causes immeasurable social and ecological damage, however, and the sudden increase in barn owls in no way compensates for the systematic destruction of biodiversity.

## FURTHER READING

Altwegg, R., Roulin, A., Kestenholz, M. and Jenni, L. 2006. Demographic effects of extreme winter weather in the barn owl. *Oecologia* **149**: 44–51.

Bell, P. J. and Mooney, N. 2002. Distribution, habitat and abundance of masked owls (*Tyto novaehollandiae*) in Tasmania. In Newton, I., Kavanagh, R., Olsen, J. and Taylor, I. (eds) *Ecology and Conservation of Owls*. Melbourne: CSIRO, pp. 125–136.

Blaker, G. B. 1933. The barn owl in England. *Bird Notes and News* **15**: 207–211.

Bruijn, O. de. 1994. Population ecology and conservation of the barn owl *Tyto alba* in farmland habitats in Liemers and Achterhoek (the Netherlands). *Ardea* **82**: 1–109.

Colvin, B. A. 1985. Common barn-owl population decline in Ohio and the relationship to agricultural trends. *J. Field Ornithol.* **56**: 224–235.

Dadam, D., Barimore, C. J., Shawyer, C. R. and Leech, D. I. 2011. The BTO barn owl monitoring programme 2000–2009. *BTO Research Report* **557**.

Huang, A. C., Elliot, J., Martin, K. and Hindmarch, S. 2013. SARA-listed (Species At Risk Act) barn owls (*Tyto alba*) in British Columbia: genetic diversity, connectivity, and divergence. Worldwide Raptor Conference, October 2013, Bariloche, Argentina.

Huffeldt, N. P., Aggerholm, I. N., Brandtberg, N. H., Jorgensen, J. H., Dichmann, K. and Sunde, P. 2012. Compounding effects on nest-site dispersal of barn owls *Tyto alba*. *Bird Study* **59**: 175–181.

Lenton, G. M. 1984. Moult of Malayan barn owls *Tyto alba*. *Ibis* **126**: 188–197.

Shepherd, C. R. and Shepherd, L. A. 2009. An emerging Asian taste for owls? Enforcement agency seizes 1,236 owls and other wildlife in Malaysia. *BirdingASIA* **11**: 85–86.

Toms, M. P., Crick, H. Q. P. and Shawyer, C. R. 2001. The status of breeding barn owls *Tyto alba* in the United Kingdom 1995–97. *Bird Study* **48**: 23–37.

Williams, V. L., Cunningham, A. B., Kemp, A.C. and Bruyns, R. K. 2014. Risks to birds traded for African traditional medicine: a quantitative assessment. *PLoS One* **9**: e105397.

Blue pellets containing acute rodenticides (1080 sodium fluoroacetate) at the entrance of a vole tunnel in Israel. © Motti Charter

## 2.4  POLLUTION

# Pernicious human practices

Owls suffer from the human use of poisons and other pollutants, just as many other organisms do. Although the situation is worrying, the barn owl's situation offers some refreshing optimism. In many parts of the world, barn owls are no longer the target of intentional poisoning, and their appetite for small mammals means that they are even used as efficient biological pest control agents.

The by-products of industrial activity are found in the air, soil and water, contaminating nature. Barn owls are no exception. Brominated flame retardants used for plastics, textiles, insulation materials, electrical and electronic equipment are found in large quantities in Dutch barn owls. Polluted air affects not only city dwellers but owls as well. Polycyclic aromatic hydrocarbons are volatile compounds and, not surprisingly, strongly affect the owl lung, as has been found in Spain.

Owls are mainly contaminated through their food, and their position at the top of the food chain is highly problematic because of bioaccumulation: poisons accumulate in the predator's body following repeated consumption of contaminated prey. Pollutants such as aluminium, arsenic and lead are found in owl feathers and pellets, but unfortunately the production of these is not enough to detoxify the body entirely. We must continue to conduct eco-toxicological studies to raise awareness of the extent of contamination and the negative impact of toxic industrial and agricultural products. In the USA, the abundance of some pollutants, such as organochlorine pesticides (e.g. DDT) and polychlorinated biphenyls (PCBs), in barn owls' bodies decreased from 1975 to 1994, but the abundance of many other pollutants, such as anticoagulant rodenticides, increased.

## Anticoagulant rodenticides

Harm is inflicted on the environment by a broad suite of human activities from which few organisms profit. Although agriculture offers new food sources for some invertebrates, birds and small mammals, many of these ultimately contribute to crop loss and are considered as pests. The farmer's way of destroying these is chemical warfare: they respond to an evil with an evil!

The first generation of anticoagulant rodenticides was introduced in the 1940s, but small mammals rapidly developed resistance. A second-generation rodenticide, 100–1000 times more toxic, was then developed in the mid-1970s. The latter compounds typically concentrate in the liver of rats and mice, leading to haemorrhage. As poisoned rodents are mobile for up to 14 days before dying, the chemicals used to control rats and mice in and around buildings may also poison predators that forage in such places. By eating poisoned small mammals, predators accumulate rodenticides in their bodies, and some ultimately die. The control of pests that ends up killing their predators is ultimately counterproductive. The snake bites its own tail!

Researchers have found rodenticide residues in predatory bird carcasses in many countries, including Canada, the Canary Islands, Denmark, France, Hungary, Spain and the United Kingdom. In some studies, barn owls are less contaminated than other bird species. In Scotland, for example, anticoagulant residues were detected in 35% of dead barn owls compared with 69% of dead red kites and 54% of dead sparrowhawks, and in Canada they were detected in 62% of dead barn owls compared with 92% of dead barred owls. However, results from the United Kingdom's Predatory Bird Monitoring Scheme have shown that the vast majority of barn owls contain one or more rodenticides. In 2015, for example, 94% of barn owls were found to contain rat poison, and in 2011, 100% of kestrels and 94% of red kites were similarly contaminated. In the vast majority of cases the level of contamination was sub-lethal. In one British study, only 12 out of 1009 barn owls found dead in the wild succumbed as a result of rodenticides. Unfortunately the effects of sub-lethal doses are unknown.

Experimental studies performed in India, England and the USA have clearly shown that barn owls die if they regularly eat mice poisoned with anticoagulant rodenticides. However, if sub-lethal contamination is affecting behaviour, survival or productivity, the full impacts of poison on barn owl populations may be far more substantial than assumed.

## Barn owls as biological pest control agents

The use of anticoagulant rodenticides is a major environmental problem. We therefore need to find an environmentally friendly solution that satisfies farmers and the consumers of agricultural products, a growing proportion of whom request organic food. The barn owl feeds almost exclusively on small mammals located in agricultural fields: the species can be locally common, and where prey is abundant each pair can produce many hungry nestlings. Indeed, a family of barn owls can consume more than 6000 voles in a single year!

In several countries, farmers and ornithologists have understood the key role played by barn owls in agriculture. In Israel, Malaysia, the USA and Venezuela large numbers of nest boxes have been erected, to favour this 'biological pest control agent' over the use of chemicals to control small mammals. In Malaysia, damage to rice fields due to rats is significantly lower when the density of nest boxes is higher, which indicates a positive effect of barn owls on agriculture. Economic analyses in Israel further suggest that the addition of barn owl boxes leads to a net economic benefit for farmers (US$235 per hectare per year), who can spread less rodenticide and simultaneously improve crop yield by 9.4%. The gain is even higher when we consider that the reduced use of rodenticides helps preserve soils and can bring public health benefits for human populations. Clearly this is a win–win solution. More data are needed to evaluate the positive impact of barn owls in combating small mammals in other countries.

## Do not introduce owls on islands

The success story of the barn owl as a biological pest control agent has inspired ecologists. They have introduced this species on several islands to combat mice and rats that were themselves introduced by humans. South African barn owls

The barn owl hunts rodents very efficiently and is therefore a perfect pest control agent.

were successfully introduced on St Helena in 1937 and the Seychelles in 1951, Californian owls in Hawaii in 1958–1963 and Australian masked owls on Lord Howe Island between 1922 and 1930. However, we should not forget that the barn owl is particularly successful in colonizing new regions and is an efficient predator. Generally speaking, biodiversity on islands is impoverished compared to that on the mainland. In situations where there are few small mammals to consume, barn owls can turn to eating endemic or rare birds, especially where they are vulnerable to predation. If the (re)-introduction of animals is necessary, it should be done using local individuals reared in breeding facilities, if possible. Furthermore, biological pest control is best performed using a predator that is already present on an island rather than a new animal whose presence might have disastrous consequences.

## FUTURE RESEARCH
- The abundance of rodents, rodent damage and crop yield should be measured in controlled experimental field trials where barn owls (1) can breed in nest boxes and (2) cannot breed in nest boxes.
- Eco-toxicological studies should be performed not only on dead owls but also on live owls.
- The effect of low levels of contamination should be explored.

Nest box monitoring in Israel. © Laurent Willenegger

## FURTHER READING

Albert, C. A., Wilson, L. K., Mineau, P., Trudeau, S. and Elliott, J. E. 2009. Anticoagulant rodenticides in three owl species from western Canada, 1988–2003. *Arch. Environ. Contam. Toxicol.* **58**: 451–459.

Ansara-Ross, T. M., Ross, M. J. and Wepener, V. 2013. The use of feathers in monitoring bioaccumulation of metals and metalloids in the South African endangered African grass-owl (*Tyto capensis*). *Ecotoxicol.* **22**: 1072–1083.

Barn Owl Trust. 2012. *Barn Owl Conservation Handbook*. Exeter: Pelagic, pp. 258–271.

Christensen, T. K., Lassen, P. and Elmeros, M. 2012. High exposure rates of anticoagulant rodenticides in predatory bird species in intensively managed landscapes in Denmark. *Arch. Environ. Contam. Toxicol.* **63**: 437–444.

Denneman, W. D. and Douben, P. E. 1993. Trace metals in primary feathers of the barn owl (*Tyto alba guttata*) in the Netherlands. *Environ. Pollut.* **82**: 301–310.

Elliot, J. E., Hindmarch, S., Albert, C. A., Emery, J., Mineau, P. and Maisonneuve, F. 2014. Exposure pathways of anticoagulant rodenticides to nontarget wildlife. *Environ. Monit. Assess.* **186**: 895–906.

Eulaers, I., Jasper, V. L. B., Pinxten, R., Covaci, A. and Eens, M. 2014. Legacy and current-use brominated flame retardants in the barn owl. *Sci. Total Environ.* **472**: 454–462.

González Amigo, S., Simal Lozano, J. and Lage Yusty, M. A. 2002. Extraction of polycyclic aromatic hydrocarbons in barn owls (*Tyto alba*) from northwest Spain by SFE and HPLC-FL analysis. *Polycyc. Aromat. Compd.* **22**: 1–11.

Hafidzi, M. N. and Na'im, M. 2003. The use of the barn owl, *Tyto alba*, to suppress rat damage in rice fields in Malaysia. *Aciar Monograph* **96**: 274–277.

Labuschagne, L., Swanepoel, L. H., Taylor, P. J., Belmain, S. R. and Keith, M. 2016. Are avian predators effective biological pest control agents for rodent pest management in agricultural systems? *Biol. Control* **101**: 94–102.

Barn owls on islands can specialize in hunting rare birds. This nest on the Alegranza archipelago in the Canary Islands is located in a hole in the ground and is surrounded by feathers of Bulwer's petrels. © Laura Gangoso

Mendenhall, V. M. and Pank, L. F. 1980. Secondary poisoning of owls by anticoagulant rodenticides. *Wilson Soc. Bull.* **8**: 311–315.

Newton, I. and Wyllie, I. 2002. Rodenticides in British barn owls. In Newton, I., Kavanagh, R., Olsen, J. and Taylor, I. (eds) *Ecology and Conservation of Owls*. Melbourne: CSIRO, pp. 286–295.

Newton, I., Wyllie, I. and Freestone, P. 1990. Rodenticides in British barn owls. *Environ. Pollut.* **68**: 101–117.

Newton, I., Wyllie, I., Gray, A. and Eadsforth, C. V. 1994. The toxicity of the rodenticide flocoumafen to barn owls and its elimination via pellets. *Pest. Sci.* **41**: 187–193.

Saravanan, K. and Kanakasabai, B. 2004. Evaluation of secondary poisoning of difethalione, a new second-generation anticoagulant rodenticide to barn owl, *Tyto alba* Hertert under captivity. *Indian J. Exp. Biol.* **42**: 1013–1016.

Shore, R. F., Walker, L. A., Potter E. D., Pereira, M. G., Sleep, D., Thompson, N. J. 2017. Second generation anticoagulant rodenticide residues in barn owls 2016. CEH contract report to the Campaign for Responsible Rodenticide Use (CRRU) UK, 21 pp.

Walker, L. A., Chaplow, J. S., Llewellyn, N. R., Pereira, M. G., Potter, E. D., Sainsbury, A. W., and Shore, R. F. 2013. Anticoagulant rodenticides in predatory birds 2011: a Predatory Bird Monitoring Scheme (PBMS) report. Lancaster: Centre for Ecology & Hydrology.

An image from the past in Europe. A barn owl crucified on a door was believed to ward off evil spirits in Christian countries. Unfortunately, in many parts of the world, barn owls are still believed to be harbingers of bad luck.

## 2.5    WHAT CAN WE DO TO PROTECT BARN OWLS?

# Joining forces to protect nature

**We must welcome any effort seeking to preserve the environment and promote sustainability. Measures applied at large geographic scales, such as restoring habitats and adding artificial nest sites, are particularly efficient in protecting barn owls. Preventive and educational action should be taken to reduce owl mortality.**

Owl populations are at risk. Major causes of population declines include loss of hunting habitats and nest sites, man-made hazards including traffic, toxic chemicals and increased human disturbance, predation, harsh winters and other extreme weather events. The list of risks is too long, and we need to deploy a number of conservation measures to help owls. Four approaches can be implemented, namely:

(1) Conservation measures targeted at populations
(2) Saving injured or starving individuals
(3) Prevention of harmful activities
(4) Education of the human population

The concept of nature conservation is broad enough to salute any effort made in the right direction to help our environment, restore or protect biodiversity, sustain populations and ensure the wellbeing of each individual. These are the humble streams that flow into great rivers.

**Targeting the population** In many regions, populations have declined dramatically. To address the situation, we can treat either the causes of such declines or the symptoms. Although these approaches are sometimes complementary, they do not always have the same value. For example, a breed-and-release programme will have little impact if it does not ensure that the released birds can breed and survive. In the 1980s, owl enthusiasts in the United Kingdom were releasing at least 3000 captive-bred owls per year. In Iowa, Missouri and Nebraska, 1000 were released annually over a period of six years. Unfortunately, these schemes had no detectable long-term effect on barn owl population levels.

The best measure for preventing and reversing owl population declines is the restoration of their foraging habitat. A return to more traditional forms of agriculture is often advised, because traditional practices benefit not just owls but other organisms as well. However, this may appear utopian, given that some economic models advise the opposite – for example, clearing rainforest for fields of soy or palm plantations. To restore environmental resilience and sustainability, ornithologists, and conservationists, must work at all levels, including in the political sphere, to convince states to implement agricultural methods that are more respectful of nature. We can also work directly with farmers to promote wildlife-friendly measures such as the re-creation of prey-rich foraging habitat and the re-planting of hedgerows that have been sacrificed in the name of agricultural intensification. This will help support not only barn owls, but a wide range of flora and fauna including the owls' prey.

A second measure, and by far the most popular, is the provision of artificial nesting sites. Barn owls adapted to utilizing traditional barns and other man-made structures for nesting thousands of years ago. Since then, many old farm buildings and many of the old hollow trees they used to inhabit have disappeared from the countryside. The conversion of old barns into human dwellings and in some cases an intolerance of sharing space with these creatures has further reduced the availability of suitable sites. A simple solution to this problem is to erect nest boxes in rural buildings, on trees, or on poles.

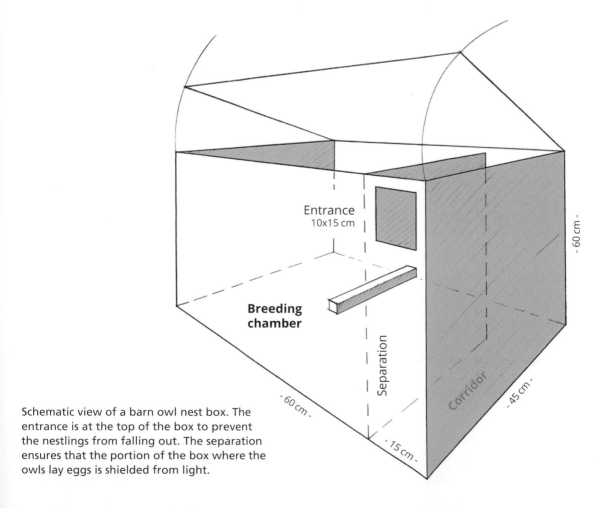

Schematic view of a barn owl nest box. The entrance is at the top of the box to prevent the nestlings from falling out. The separation ensures that the portion of the box where the owls lay eggs is shielded from light.

In regions where site availability is a limiting factor, the provision of nest boxes over wide areas can help populations expand through immigration from neighbouring 'source' populations. In very hot climates, boxes should be placed in shady areas to prevent eggs and owlets from overheating.

In the United Kingdom, an estimated 25 000 barn owl nest boxes had already been erected by 1997, and since then many thousands more have been provided – but there has been little benefit across most of the country, where food availability is the limiting factor. However, in Denmark, the erection of nest boxes resulted in a 20-fold increase in population size between 1998 (25 pairs) and 2009 (500 pairs). Similar results have been obtained in Germany and the Czech Republic, where there are more than 20 000 and 4 000 nest boxes, respectively.

**Saving injured or starving individuals**  Another approach is to consider individuals rather than populations as the unit of protective intervention. Although such actions are unlikely to have a significant impact on barn owl populations as a whole, these efforts demonstrate that our society respects animals. While ecologists are more concerned with the fate of animal populations or species as a whole, the lay public is particularly sensitive to the vision of suffering and death.

# Barn owl nest boxes: a fact sheet

- Nest boxes should be erected to compensate for the loss of historic roosts and nest sites. Because barn owls often change sites to produce a second annual brood, nest boxes are often erected in pairs.
- Boxes should be sufficiently large (e.g. 60 × 60 × 45 centimetres) to host large families and sufficiently deep so the nestlings do not fall to the ground (minimum 45 centimetres from the bottom of the hole to the bottom of the box).
- Boxes are usually made of wood, but in countries where wood is scarce, other material should be used. It is also possible to recycle boxes, such as in Jordan, where ballot boxes are used as barn owl nest boxes!
- Because owls do not bring any material to their nest, a dry organic substance such as wood flakes should be added.
- Pairs of nest boxes are usually placed at least 300 metres apart. However, barn owls are not always territorial and where food is plentiful nests can be very close together.
- Nest boxes can be installed in isolated barns. However, in some countries boxes are usually erected in church towers in villages, to enable the barn owls to evade nest predation by martens. Nest boxes can also be installed in busy or noisy places, as barn owls that are able to hide can learn to tolerate loud noises. Provision for barn owls should be incorporated into rural barn conversions. Provided this is done well, there are no health or nuisance implications.
- Nest boxes are normally installed not less than 3 metres above the ground. However, where martens are likely to be a threat, they should be fixed at a height of at least 5 metres. It may also be necessary to take extra steps to prevent predator access.
- In very hot climates, nest boxes can have a double wall and roof, and additional ventilation, to keep the inside of the box from overheating. In addition, when fixing a box inside a barn, one should avoid fixing it at the highest point where ambient temperature may be higher.
- Owls like to remain in dark places during daylight hours. A separation between the nest entrance and the place where the owls lay the eggs is recommended, to ensure that a large portion of a nest box remains in the dark.
- For further information see the *Barn Owl Conservation Handbook* or visit www.barnowltrust .org.uk.

In the Middle East, nest boxes are often fixed on poles in fields.

Nest boxes can be fixed against the walls of barns, with the box being either outside or inside the building. When inside, a hole is pierced in the wall so that the owls enter the box from the outside.

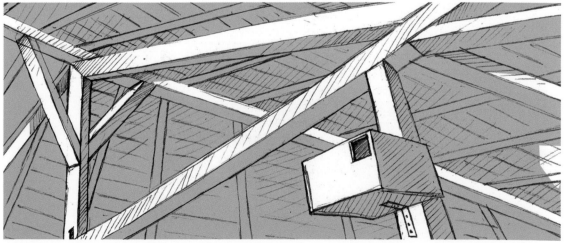

Nest boxes can be fixed inside barns, although in this situation there is a risk that martens will reach barn owl nests.

Barn owls are fascinating and can be a flagship species for nature conservation.

In many countries, nature lovers, and even governments, have created rehabilitation centres to treat injured and food-deprived animals. Large numbers of individuals can be saved in this way, as shown in Germany, where treated barn owls have a high survival rate once released into the wild. Because a major cause of mortality is starvation, rehabilitation centres in Spain have encouraged released barn owls to remain nearby by offering shelter and food. The use of supported or 'soft' release methods substantially enhances their survival. However, individuals that have suffered serious trauma, such as bone fractures, have less chance of survival once released – and for them a supported release method is equally important. Individuals that never fully recover fare much better when kept in captivity rather than left to their fate in the wild. Captive disabled birds can sometimes be used in educational programmes. However, they never become tame and tend to suffer stress. Most owls used for educational purposes are captive-bred and hand-reared.

Straightforward, simple actions can save individual birds. Ornithologists also sometimes have the opportunity to reunite families or create new families. When food supply is momentarily low – owing, for instance, to inclement weather that inhibits foraging – hungry offspring try to attract the attention of their parents by remaining close to the entrance of the nest cavity and calling. When a parent finally arrives with food, nestlings sometimes push each other in an attempt to get the food, and one may accidentally fall out of the nest. Because parents do not feed their offspring outside the nest, the fallen individual will starve unless returned to its siblings. When a nest is destroyed or a nestling is orphaned, ornithologists can transfer young to other nests to be adopted by parents who are apparently unable to distinguish their own offspring from others.

**Prevention of harmful activities** Most causes of mortality are known, and simple measures can sometimes prevent them. For instance, nestlings can be prevented from falling by the provision of deep nest boxes. Cattle often drink from large water tanks placed in fields. When thirsty owls land on the edge of these tanks to drink or to take a bath, they frequently drown. A simple solution to this problem is to install a wooden float that can support the weight of an owl but still allows the cattle to drink.

Owls are also often trapped inside abandoned buildings: while it may be easy to find an entrance, it is sometimes more difficult to find an exit! Leaving a window wide open would prevent owls from starving to death. Additionally, owls fly close to the ground and thus often collide with cars and trucks. To reduce the risk of collision, roads would be safer if placed lower than the surrounding fields or screened with trees and shrubs to force the owls to fly above the height of most vehicles. Electricity supply cables on poles can be insulated to prevent birds being electrocuted. Unwanted rodent populations can be controlled by environmental management rather than the use of highly toxic poisons. Starvation can be reduced by the provision of prey-rich habitat, and nest mortality can be reduced by providing safe nesting places inside rural buildings.

**Fascination** Owls have long fascinated humans, for good and bad reasons. There are still countries where people capture owls from their nests and sell them in markets as pets. Owls also stimulate human imagination. Perhaps one of the most famous recent examples is the role played by owls, particularly the snowy owl, in movies such as the Harry Potter series (which also caused an increase in owls being caught for the pet trade). Conservationists should capitalize on this fascination and harness public support for their conservation.

**Education** The power of education to change perceptions is undeniable. In the past, European farmers nailed (or crucified) barn owls against the walls of their farms to ward off evil spirits. Crucifying owls was so frequent in the United Kingdom that this bird was known as the 'barn door owl'! Nowadays, in stark comparison, awareness is spreading in the farming community and some farmers proudly invite their friends to see how cute barn owl chicks are. Nevertheless, in some countries, barn owls are still

A barn owl parent bringing a wood mouse to its nest in France. © Alex Labhardt

Barn owls often hunt by flying just above the ground, and hence are at risk of collision with cars.

considered bad omens. But it is only a matter of time before people realize that these birds help farmers by eating mice, rats and voles. Education about the importance of owls in maintaining a balanced ecosystem is essential. Unfortunately, other powerful interests are at play – agrochemical industries encourage farmers to use their chemical products and may see a move towards nature-friendly methods as a threat to their profits.

## FURTHER READING

Bairlein, F. and Harms, U. 1994. Ortsbewegungen, Sterblichkeit und Todeursachen von Greifvögeln und Eulen nach Ringfunden der 'Vogelwarte Helgoland' – eine übersicht. *Vogelwarte* **37**: 237–246.

Barn Owl Trust. 2012. *Barn Owl Conservation Handbook*. Exeter: Pelagic.

Colvin, B. A. 1981. Nest transfer of young barn owls. *Ohio J. Sci.* **81**: 132–134.

Fajardo, I., Babiloni, G. and Miranda, Y. 2000. Rehabilitated and wild barn owls (*Tyto alba*): dispersal, life expectancy and mortality in Spain. *Biol. Conserv.* **94**: 287–295.

Graves, G. R. 2016. Observation of a barn owl (*Tyto alba*) bathing in a rainwater pool in Jamaica. *J. Caribb. Ornithol.* **29**: 16–17.

Nijman, V. and Nekaris, K. A.-I. 2017. The Harry Potter effect: the rise in trade of owls as pets in Java and Bali, Indonesia. *Global Ecol. Conserv.* **11**: 84–94.

# 3 Parasites and predators

3.1   Endoparasites
3.2   Ectoparasites
3.3   Predators and anti-predator behaviour

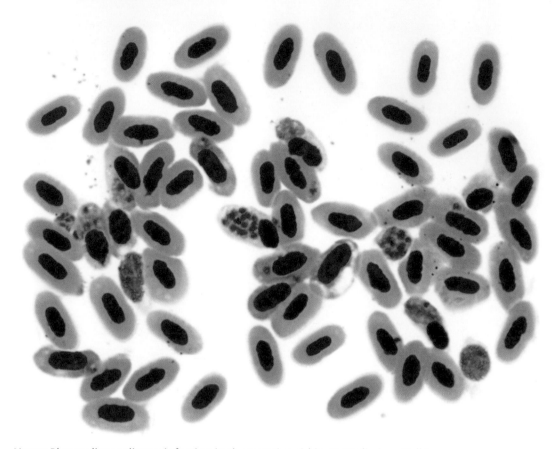

Heavy *Plasmodium relictum* infection in the Eurasian siskin. © Gediminas Valkiunas

## 3.1  ENDOPARASITES

# From the inside

**Endoparasites attack organisms from within. They have tremendous effects on their host and, in turn, on ecosystems, but little is known about endoparasitism in the barn owl. Considering the mode of parasite transmission provides insight into the barn owls' risks of contracting diseases.**

Barn owls have evolved a battery of behavioural, morphological and physiological adaptations to resist parasite attack. Although these organisms withdraw resources from their avian hosts and are responsible for many diseases, parasites should not *a priori* be responsible for decreases in barn owl population sizes. Parasites and their hosts have co-evolved over thousands of years and should be adapted to each other. Hence, many parasites are not too virulent, so as not to kill their host, and hosts can handle parasites without substantial loss in fertility or survival capacity.

Unfortunately, however, the threat of disease may suddenly become real when human activities favour the spread of novel exotic parasites to which their hosts have not yet evolved resistance strategies. This has happened to many animals in remote parts of the world, such as the Galápagos Islands.

Although the exact impact of endoparasites exploiting barn owls remains unclear, we know that intensive agricultural practices and human disturbance increase physiological stress, as found in Swiss barn owls. This is something that can weaken the host immune system and hence the ability to cope with parasites. Monitoring endoparasites is therefore of key importance to understand their potential impact on owl populations. This goal, however, has not been met. Only a handful of researchers have determined the prevalence of endoparasites in Tytonidae. In South Africa 31% of barn owls and 7% of grass owls contract parasitic worms; in the Czech Republic, 13% are infested with the protozoan *Isospora buteonis*; in France, 11% are infested with *Toxoplasma gondii*; and in Ontario, 80% of barn owls have antibodies against the West Nile virus.

## Mode of disease transmission
Knowing the mode of transmission of endoparasites provides information about the risk incurred by barn owls of contracting intracellular parasites (viruses, bacteria, protozoa), which inhabit cells in the bird's body, and intercellular parasites (roundworms, flatworms, thorny-headed worms), inhabiting body cavities. All of the endoparasites mentioned below have been found in the barn owl.

Countries where intercellular endoparasites have been identified in barn owls and their relatives.

| Category of endoparasites | Species of endoparasite | Country |
|---|---|---|
| Roundworms (Nematoda) | *Baylisascaris procyonis* | North America |
| | *Capillaria tenuissima* | Netherlands, Spain, UK |
| | *Cyathostoma americana* | Netherlands, Spain |
| | *Dispharynx affinis* | USA |
| | *Dispharynx nasuta* | Italy |
| | *Gongylonema neoplasticum* | UK |
| | *Habronema murrayi* | Africa |
| | *Porrocaecum angusticollae* | Netherlands, Spain |
| | *Porrocaecum depressum* | Czech Republic, UK, USA |
| | *Porrocaecum spiralis* | Czech Republic, UK |
| | *Procyrnea leptoptera* | Spain |
| | *Syngamus trachea* | Nigeria |
| | *Synhimantus affinis* | Italy, UK |
| | *Synhimantus laticeps* | Argentina, Italy, Netherlands, Spain, UK |
| | *Trichinella pseudospiralis* | Tasmania |
| Flatworms (Platyhelminthes) | *Brachylaima fuscatum* | Italy |
| | *Brachylaima* spp. | Spain |
| | *Brachylecithum lobatum* | Czech Republic |
| | *Choanotaenia strigium* | Czech Republic |
| | *Cladotaenia globifera* | Czech Republic |
| | *Neodiplostomum americanum* | USA |
| | *Neodiplostomum attenuatum* | Spain |
| | *Neodiplostomum japonicum* | Italy |
| | *Paruternia candelabraria* | Spain, UK |
| | *Stomylotrema chabaudi* | Africa |
| | *Strigea falconis* | Czech Republic |
| | *Strigea strigis* | Czech Republic, Netherlands, UK |
| Thorny-headed worms (Acanthocephala) | *Centrorhynchus aluconis* | Italy, Slovakia, UK |
| | *Centrorhynchus conspectus* | Czech Republic, UK |
| | *Centrorhynchus globocandatus* | Czech Republic, Spain, UK |
| | *Centrorhynchus tumidulus* | UK |
| | *Porrorchis tyto* | Vietnam |

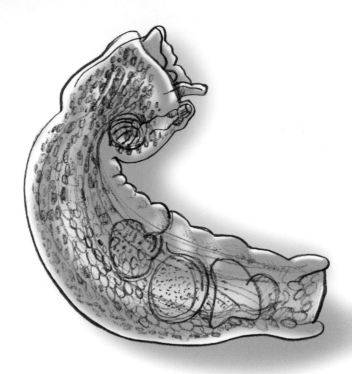

The flatworm *Strigea* sp.

By biting owls, infected invertebrates, such as mosquitoes, beetles and ticks, transmit the Saint Louis encephalitis and West Nile viruses, as well as the protozoa *Leucocytozoon ziemanni*, *L. danilewskyi*, *L. californicus*, *Haemoproteus*, *Plasmodium* and *Trypanosoma*. Although feathers cover most of their body, owls can be bitten around the bill and on their unfeathered feet. Owls living close to wetlands, where these vectors abound, may therefore be at a higher risk of contracting such vector-borne diseases.

Barn owls contract airborne bacterial diseases by inhaling infected air. These can include infections by *Salmonella typhimurium*, *S. thompson*, *S. tuindorp*, *S. choleraesuis* and *Chlamydia psittaci*, which can be transmitted from birds to humans.

Many diseases are contracted through the ingestion of faecal substances and other animal excretions present in food, water and soil. Strict hygiene is the best way to avoid such infections, but this is far from the situation in crowded barn owl nests, where nestlings and parents defecate and expel pellets within the nest cavity. This is probably the reason why barn owl and grass owl mothers and nestlings clean their nest by removing pellets and other dirty material, to decrease the risk of contracting diseases. It would be more efficient to defecate outside the nest, but in the cavity-nesting barn owl, getting too close to the nest entrance incurs the risk of falling out – something that often happens to young nestlings, which have poor locomotor balance. The lack of strict hygiene favours the spread of the protozoa *Toxoplasma gondii*, *Cryptosporidium*, *Isospora buteonis* and *Giardia*, as well as poxvirus and avulavirus. An avulavirus is responsible for Newcastle disease, which generates respiratory and digestive problems followed by nervous-system symptoms. This disease is particularly prevalent in poultry, and hence by breeding in farms barn owls may often come into contact with poultry and, in turn, Newcastle disease. Another disease of high concern is tuberculosis, caused by *Mycobacterium* species, particularly *M. avium*.

Rats can transmit pathogens such as the bacterium *Pasteurella* by biting other animals. However, are small mammals quick enough to bite barn owls before being killed? Although no information is available to answer this question, we have never observed barn owls with injuries caused by small mammals. More problematic is the consumption of prey infected with the protozoa *Trichomonas gallinae* and *Sarcocystis dispersa*, and the roundworm *Baylisascaris*.

**Intercellular parasites** Intercellular parasites inhabiting cavities in the barn owl's body include fifteen species of roundworms or nematodes, twelve species of flatworms or platyhelminths and five species of thorny-headed worms or acanthocephalans.

## FUTURE RESEARCH
- Many bacteria and viruses are contracted through the air and faecal substances. Do barn owls contract diseases by living in a dirty nest where chicks defecate and expel their pellets? Barn owls may therefore be expected to display high resistance to pathogens. Thus, we should investigate whether or not owls have evolved an efficient immune system to combat endoparasites.
- Droppings inside the nest may promote the spread of diseases. Thus, the potential effect of parental nest-cleaning behaviour on the spread of pathogens should be formally examined. However, in some nests, the ammonia resulting from droppings may kill rather than facilitate the spread of pathogens and hence play a prophylactic role. This possibility could be experimentally tested by randomly allocating nests to two groups in which extra ammonia or water (as a control) is spread, with the prediction that parasites will be less numerous in ammonia-treated nests than in control nests. Whether the level of ammonia observed in barn owl nests is poisonous to nestlings should also be tested.
- Prey is stored in the nest along with the nestlings, which may also increase the risk of contracting diseases. In hot summers, prey rots rapidly, which may favour the spread of disease, an effect that is probably amplified when nestlings consume this decaying food.
- Data are required on the prevalence and virulence of endoparasites and their impact on barn owl population dynamics. This should be studied in barn owls located in different regions, such as in the tropics versus temperate regions, or on islands versus the mainland.

## FURTHER READING
Daszk, P., Cunnigham, A. A. and Hyatt, A. D. 2000. Emerging infectious diseases of wildlife: threats to biodiversity and human health. *Science* **287**: 443–449.

Barn owl feather damaged by chewing lice.

## 3.2  ECTOPARASITES

# From the outside

**Ectoparasites transmit diseases and endoparasites to their avian hosts and consume blood, flesh and feathers. Barn owls are infected by a wide diversity of organisms, including mites, ticks, bugs, flies and various lice. Although several factors can explain why individuals are not equally affected by parasites, much more research is necessary to understand the impact of ectoparasites on barn owl breeding biology and populations.**

Ectoparasites live on the host's external body parts. Even though almost nothing is known about their impact on barn owl reproduction and survival, some information is available on the impressive parasite control strategies the barn owl has evolved.

**Feather lice and ticks**  Feather-eating lice can damage a bird's plumage by making holes, but without fingers, an owl has no way of removing them other than to scratch at them with bill and claws. A small hook on the beak, and many teeth on the serrated edge of the middle claw, a rare adaptation in birds, are particularly effective at preening plumage. Nestlings also often preen each other's plumage in so-called allopreening, possibly to kill parasites located on their siblings. Accordingly, 78% of the allopreened body parts are those that individuals cannot preen by themselves – the head, back and neck.

The removal of ticks located on bare body parts is more difficult if not impossible. The best way to resist ticks may therefore be an efficient immune system, something that requires further study.

Bill hook, and teeth on the serrated edge of the middle claw.

Unable to jump or fly, ticks drop from a branch onto their host or sit on a blade of grass, leaf or twig waiting for a passing host to come within reach. In Switzerland, ticks were found on nestlings in 19% of 330 nests, and in nests with ticks, 1.7 ticks (*Ixodes ricinus*) were detected on average in each nest. Located on the head of barn owl adults, ticks detach and crawl onto the first nestling they meet. This random process probably explains why ticks are found with equal frequency on male and female nestlings and on junior and senior nestlings.

## A blood-sucking fly *Carnus hemapterus* is a generalist ectoparasite that attacks the nestlings

of many bird species. Although it is widely distributed geographically, this species has been studied only in the USA and Switzerland, where 90–97% of barn owl nestlings are infected with an average of 39 flies, the maximum being 383 flies recorded on a single nestling. Flies lay their eggs in birds' nests, and because cavity-nesting birds such as the barn owl often reuse the same cavity year after year, the emerging flies easily find their hosts. When a family of birds does not reuse a cavity, the emerging flies colonize new avian nests by flying, and once a nest is found, the flies lose their wings. Thus, nestling barn owls are parasitized by many more flies if their nest hosted a bird family the year before than if the cavity is used for the first time or if the nest box has been cleaned between breeding seasons.

Although some flies emerge just before nestlings hatch, they avoid unfeathered hatchlings. Hence, during the first days of life when flies are still relatively rare, the first-born owlets that have already developed some feathers are more heavily parasitized than are their last-born naked siblings. As flies continue to emerge, the fly population size progressively increases until the first-born owlets become too feathered for flies. At this point, the flies change strategy and concentrate their attacks on the last-born chicks, which still have relatively few feathers. This results in the youngest nestlings harbouring most of the flies in the nest – not necessarily because they are weaker and less immunocompetent than their older siblings, but because at the time of greatest fly abundance, they are optimally feathered to host them.

To help the youngest nestlings fight this aggression, it seems that mothers add unknown substances to the later eggs to enhance resistance to parasites. However, the anti-parasitic effect of these substances is but one side of the coin, as they also appear to impair growth rate. This would explain why these substances are either not added or added at a lower concentration in the first-laid eggs. This was concluded from an experiment where eggs were exchanged between nests, allowing us to compare the number of *C. hemapterus* individuals later harboured by nestlings in relation to

The parasitic fly *Carnus hemapterus*.

the rank at which the egg had been laid in the nest of origin, and the rank of the nestlings that were raised in foster nests. Both the hatching and rearing ranks explained the abundance of ectoparasites on the nestlings.

Differences in response to parasites are also observed between males and females. For unknown reasons, female nestlings in Switzerland have slightly more flies on their body than do males (44 vs. 38 flies, on average). More surprising is the finding that nestlings are better able to resist attacks of *C. hemapterus* if their biological mother displays larger black feather spots on the tips of her ventral body feathers. It is likely that genes inherited from large-spotted mothers enhance parasite resistance.

A tick of the genus *Ixodes*.

Ectoparasites that have been recorded on barn owls.

| Mites | | | Ticks | |
|---|---|---|---|---|
| | *Aureliania aureliani* | *Noeboydaia aureliani* | | *Argas reflexus* |
| | *Bubophilus ascalaphus* | *Neottialges evansi* | | *Ixodes arboricola* |
| | *Bubophilus tytonus* | *Ornithonyssus bursa* | | *Ixodes ricinus* |
| | *Dermonoton parallelus* | *Ornithonyssus sylviarum* | | |
| | *Dermonoton sclerourus* | *Pandalura strigisoti* | **Flies** | *Carnus hemapterus* |
| | *Glaucalges attenuatus* | *Protalges attenuates* | | |
| | *Harpyrhynchus tyto* | *Rhinoecius tytonis* | **Sucking lice** | *Hoplopleura acanthopus* |
| | *Kramerella lunulata* | *Suidasia* spp. | | |
| | *Kramerella lyra* | Sarcoptiformes | **Chewing lice** | *Tytoniella rostrata* |
| | *Kramerella quadrata* | *Tytodectes striges* | | |
| | *Leptotrombidium nissani* | *Tytodectes tyto* | **Feather lice** | *Colpocephalum subpachygaster* |
| | | | | *Colpocephalum turbinatum* |
| **Bugs** | *Haematosiphon inodorus* | | | *Glaucalges tytonis* |
| | | | | *Kurodaia subpachygaster* |
| **Fleas** | *Ceratophyllus gallinae* | | | *Philopterus rostratus* |
| | *Ceratophyllus rossitensis* | | | *Strigiphilus aitkeni* |
| | *Malaraeus telchinus* | | | *Strigiphilus cursitans* |
| | *Thrassis* spp. | | | *Strigiphilus rostratus* |

**Ectoparasites recorded on barn owls** Some of the many ectoparasites that have been found on barn owls are listed in the table above. Mites can feed on blood, flesh or detritus. Many other ectoparasites are blood feeding, such as bugs, fleas, sucking lice, ticks and the fly *Carnus hemapterus*. Chewing lice eat skin and feather debris, and feather lice consume feathers.

## FUTURE RESEARCH
• Studies are required on ectoparasite prevalence and intensity at the global scale.
• The effect of ectoparasites on barn owl reproduction and survival must be investigated.

## FURTHER READING
Bush, S. E., Villa, S. M., Boves, T. J., Brewer, D. and Belthoff, J. R. 2012. Influence of bill and foot morphology on the ectoparasites of barn owls. *J. Parasitol.* **98**: 256–261.

Christe, P., Møller, A. P. and de Lope, F. 1998. Immunocompetence and nestling survival in the house martin: the tasty chick hypothesis. *Oikos* **83**: 175–179.

Clayton, D. H., Koop, J. A. H., Harbison, C. W., Moyer, B. R. and Bush, S. E. 2010. How birds combat ectoparasites. *Open Ornithol. J.* **2010**: 41–71.

Roulin, A. 1998. Cycle de reproduction et abondance du diptère parasite *Carnus hemapterus* dans les nichées de chouettes effraies *Tyto alba*. *Alauda* **66**: 1–8.

Roulin, A. 1999. Fécondité de la mouche *Carnus hemapterus*, parasite des jeunes chouettes effraies (*Tyto alba*). *Alauda* **67**: 205–212.

Roulin, A., Brinkhof, M. W. G., Bize, P., Richner, H., Jungi, T. W., Bavoux, C., Boileau, N. and Burneleau, G. 2003. Which chick is tasty to parasites? The importance of host immunology versus parasite life history. *J. Anim. Ecol.* **72**: 75–81.

Roulin, A., Christe, P., Dijkstra, C., Ducrest, A.-L. and Jungi, T. W. 2007. Origin-related, environmental, sex and age determinants of immunocompetence, susceptibility to ectoparasites and disease symptoms in the barn owl. *Biol. J. Linn. Soc.* **90**: 703–718.

Roulin, A., Gasparini, J. and Froissart, L. 2008. Pre-hatching maternal effects and the tasty chick hypothesis. *Evol. Ecol. Res.* **10**: 463–473.

Roulin, A., des Monstiers, B., Ifrid, E., Da Silva, A., Genzoni, E. and Dreiss, A. N. 2016. Reciprocal preening and food sharing in colour polymorphic nestling barn owls. *J. Evol. Biol.* **29**: 380–394.

The massive eagle owl is a major predator of barn owls.

## 3.3  PREDATORS AND ANTI-PREDATOR BEHAVIOUR

# An eye for an eye, a tooth for a tooth

**Displaying a conspicuous white colouration and emitting loud begging calls increase the risk of predation. Barn owls often fall prey to mammalian predators, larger owls and even diurnal birds of prey. To reduce predation risk, nestlings hiss like snakes to frighten potential predators.**

The quote from C. J. Ellisson, '*There's always someone bigger and badder who can knock you off your perch*', could have been written for the barn owl. Although it is extremely efficient at capturing small mammals, the relatively small size of the barn owl, its conspicuous white plumage and the loud begging calls of its chicks make it vulnerable to larger predators. While nestlings are particularly susceptible to predation, adults are also at risk when flying across open landscapes.

In Bulgaria, a single pair of eagle owls consumed at least 20 barn owls over 12 years. Ornithologists have reported avian predation on African grass owls by the martial eagle and African marsh harrier, and on barn owls by the eagle owl (64 cases), great horned owl (13), giant eagle owl (1), tawny owl (2), peregrine falcon (8), prairie falcon (1), lanner falcon (1), common buzzard (9), ferruginous hawk (2), golden eagle (2), goshawk (14), red kite (1) and hen harrier (1).

In Europe, beech martens occupy similar human-made sites as do barn owls. Although these martens can often easily reach barn owl nests, predation exerted by mustelids does not have a strong

Eagle owl pellet containing a barn owl leg.

impact on barn owl populations. In France, beech martens visited only 61 out of 951 churches occupied by barn owls (6.4%), and out of 1031 broods, only 7 (0.7%) were preyed upon. While predation is not the primary cause of barn owl mortality in the northern hemisphere, the situation in the southern hemisphere is different. For instance, the African grass owl breeds on the ground, making it easy for mammals such as black-backed jackals to reach the nest.

Interspecific interactions other than predator–prey interactions are rarely observed. However, in the British Isles, barn owls frequently fly during daylight hours with a prey in their talons, and kestrels, and even hobbies, can be seeing chasing them to steal their prey. At night, Indian flying foxes and naked-rumped tomb bats can mob barn owls. It remains unknown whether this behaviour is very rare, or simply infrequently reported because it is difficult to observe.

### Relationships between adult barn owls and humans
Birds can display high aggression towards humans when they are perceived as potential predators. For instance, the famous photographer Eric Hosking lost one of his eyes to a female tawny owl while taking pictures of her nest. However, only two cases of barn owls attacking humans have been reported, in Australia and Germany. In fact, we can often inspect nest contents by gently lifting up an incubating female without any reaction. Masked owls might be more aggressive towards humans, as seen in captivity. Why do barn owls not take more risks to protect their young against predators? This is perhaps because the barn owl is much smaller and lighter than the more powerful masked owl, making attacking a predator far riskier.

### Nestlings frighten or attack their predators
While relatively harmless, nestlings of barn owls, masked owls and grass owls can raise their claws against their predators and often produce scary snake-like hissing calls, making them quite impressive. In the presence of humans, nestlings start to breathe loudly, and as the risk of predation increases, they click their tongues, snap their beaks and hiss. Hissing calls are impressive, lasting between 1.8 and 10 seconds, and older nestlings are scarier not only because they are bigger but also because they produce longer hissing calls than do younger nestlings. Interestingly, the individual that hisses the most in a brood tends to be the most docile when handled. This means that nestlings either pretend to be dangerous by producing loud snake-like hissing calls or attack predators without warning!

Being on constant alert to detect predators requires acute selective attention to discriminate between the noise of a predator and irrelevant stimuli, such as the noise made by an arriving parent or the wind. These errors first occur at two weeks of age and are most frequent at three weeks, when nestlings produce hissing calls at up to 40% of the parental visits. Older nestlings are less likely to make these errors, probably because they have got better at identifying the source of the noise. Regardless of age, siblings often differ in the likelihood of hissing when a sound is heard. Sharing a nest with siblings that quickly hiss after having heard a noise certainly reduces the risk of predation; however, once the danger has disappeared, these individuals have difficulty calming down and continue to breathe loudly, click the tongue or hiss. This noisy behaviour scares their brothers and sisters, who consequently resume hissing behaviour. The time required for the entire family to cool down can be quite long, which adds to the stress caused by predators. While hissing behaviour is important to avoid predation, it also entails non-negligible physiological costs.

## FUTURE RESEARCH
- Anti-predator behaviour in barn owl populations where the risk of predation differs should be assessed. Of particular interest are owl populations located in tropical regions, where predators may be more abundant and diverse than in temperate regions.
- The role of predators in barn owl population dynamics should be investigated.

## FURTHER READING
Van den Brink, V., Dolivo, V., Falourd, X., Dreiss, A. N. and Roulin, A. 2012. Melanic color-dependent antipredator behavior strategies in barn owl nestlings. *Behav. Ecol.* **23**: 473–480.

# 4 Physiology
## in an ecological context

4.1   Hearing capacity
4.2   Visual capacity
4.3   Daily food requirements
4.4   Pellet production
4.5   Capacity to withstand cold weather

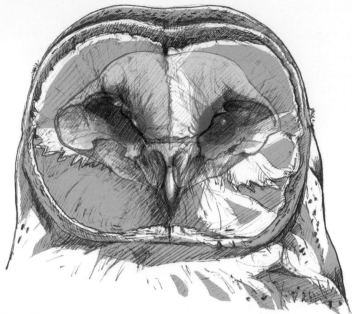

The barn owl's ruff.

## 4.1  HEARING CAPACITY

# Absolute pitch

**Dim light has promoted the evolution of sharp hearing in nocturnal predators, enabling them to forage effectively in the dark. Barn owls can hear sounds that are ten times fainter than the faintest sounds humans can detect.**

The barn owl and its relatives are ideal organisms for studying the neurophysiology of hearing. Their acute sense of hearing allows them to locate the source of a sound within 1.6–3 degrees of precision even in the total absence of light. In this regard, masked owls are even more precise than diurnal hunters such as swamp harriers and brown falcons. In addition, once a sound has reached the bird's ears, this information is rapidly processed, with the pupil dilating in only 25 milliseconds. Even though owls can hunt in complete darkness, vision still improves performance under the dim light conditions that they usually experience.

Compared to other nocturnal birds, the barn owl is particularly efficient at detecting all sound frequencies, owing to a number of anatomical and physiological adaptations. Of all the birds that have been investigated, the barn owl has the highest number of hair cells for detecting sound and the longest cochlea and auditory portion of the inner ear. A particularly well-developed morphological adaptation is a prominent facial disc that in effect increases head size. This adaptation may also have evolved in birds such as harriers, which hunt prey hidden in the grass while flying at low speeds a few metres above the ground. The sound box formed by the facial disc measures approximately 35 square centimetres, works like a satellite dish and permits amplification by up to 20 decibels of the sounds travelling to the owl's asymmetrically arranged ears. One ear is placed slightly higher in the skull than the other, and this asymmetry generates an additional difference in level and timing when a sound reaches the left and right ears, which greatly facilitates sound localization, particularly at low frequencies.

Symmetrical ears produce an **interaural time difference** that allows birds to determine the azimuthal direction of sound emission – in other words, the origin of the sound in the horizontal plane. This is

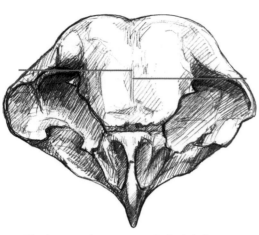
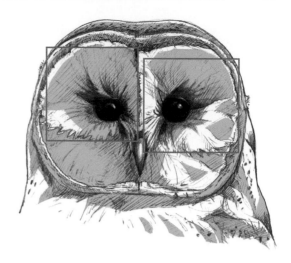

The barn owl's asymmetric facial disc.

how most animals hear, with the body simply attenuating the sound on the side that is opposite to the sound source, generating a difference between the left and right ears. The asymmetry in the elevation of the barn owl's ears causes an additional time and – more importantly – level difference (**interaural level difference**) that allows for precise localization of the emitted sound in the vertical plane. Even the size of the ear openings is different. Asymmetric ears are typically found in nocturnal owls, allowing them to accurately locate sound in both the horizontal and vertical planes.

## FUTURE RESEARCH
- Three non-mutually exclusive hypotheses may explain why natural selection favoured a very efficient hearing capacity. These hypotheses are based on the idea that in order to survive, barn owls must be exceptional hunters: (1) parents must meet the needs of their large broods; (2) for their size, barn owls hunt small prey, implying that they must capture many small prey items rather than a few large items like larger predators; and (3) barn owls exploit habitats that require exceptional hearing for hunting.
- An exceptional hearing capacity may be a problem for detecting prey in a noisy environment, such as during rainy or windy nights. This may partly explain why barn owls avoid hunting when it is raining.

## FURTHER READING
Bala, A. D. S. and Takahashi, T. T. 2000. Pupillary dilation response as an indicator of auditory discrimination in the barn owl. *J. Comp. Physiol. A* **186**: 425–434.

Dyson, M. L., Klump, G. M. and Gauger, B. 1998. Absolute hearing thresholds and critical masking ratios in the European barn owl: a comparison with other owls. *J. Comp. Physiol. A* **182**: 695–702.

Hausmann, L., Platchta, D. T. T., Singheiser, M., Brill, S. and Wagner, H. 2008. In-flight corrections in free-flying barn owls (*Tyto alba*) during sound localization tasks. *J. Exp. Biol.* **211**: 2976–2988.

Hausmann, L., von Campenhausen, M., Endler, F., Singheiser, M. and Wagner, H. 2009. Improvements of sound localization abilities by the facial ruff of the barn owl (*Tyto alba*) as demonstrated by virtual ruff. *PLoS One* **4**: e7721.

Knudsen, E. I., Blasdel, G. G. and Konishi, M. 1979. Sound localization by the barn owl (*Tyto alba*) measured with the search coil technique. *J. Comp. Physiol. A* **133**: 1–11.

Konishi, M. 1973. Locatable and nonlocatable acoustic signals for barn owls. *Am. Nat.* **107**: 775–785.

Payne, R. S. 1971. Acoustic location of prey by barn owls (*Tyto alba*). *J. Exp. Biol.* **54**: 535–573.

Peciscs, T., Laczi, M., Nagy, G., Kondor, T. and Csörgö, T. 2018. Analysis of morphometric characters in owls (Strigiformes). *Ornis Hung.* **26**: 41–53.

Von Campenhausen, M. and Wagner, H. 2006. Influence of the facial ruff on the sound-receiving characteristics of the barn owl's ears. *J. Comp. Physiol. A* **192**: 1073–1082.

The large eyes of an oriental bay owl in Jianfengling, Hainan, China. © Robert Hutchinson

## 4.2  VISUAL CAPACITY

# In the land of the blind, the one-eyed owl is king

**Although nocturnal organisms have typically evolved high-quality optics to enhance acuity under dim light, the exceptional hearing capacity of the barn owl may be sufficient for hunting. Indeed, its visual acuity and contrast sensitivity are not the best among animals, particularly under daylight conditions.**

In bright light, the visual acuity of the human, falcon, pigeon, tawny owl and great horned owl surpasses that of the barn owl. However, the nocturnal acuity of the barn owl, which is as good as its diurnal acuity, outcompetes that of the cat. Large objects are therefore distinctly seen both during the day and at night.

**Physiological adaptations to nocturnal life**  Nocturnal life has necessitated the evolution of visual adaptations, but everything has a limit:

- Large eyes with a long tubular shape, providing long focal lengths and big lenses. The eyes are 'squeezed' into rigid bones to protect them and to prevent them from popping out of the owl's skull. While a round-shaped eyeball has many degrees of freedom, a tubular eye is more restricted and cannot move easily. This also explains why barn owl eyes are almost completely immobile in their sockets.
- Adapting to the detection of low light levels, for which light-sensitive rods are required, shifts the balance away from the colour-vision cone cells because of the limited space on the retina. The ratio of the number of cells specialized to detect objects under low light intensities (rods) to the number of cells that detect colour differences (cones) is twice as large in barn owls (20:1) as in humans (10:1).
- Evaluating the distance from moving prey is easier with stereo vision, which is achieved when the fields of view of the two eyes overlap. The overlap is 42 degrees in the barn owl, compared with 20 degrees in the pigeon. This is possible because the owl's eyes are positioned anteriorly, and close to each other, rather than laterally as in the pigeon. Unfortunately, this adaptation restricts

the overall field of view (200 degrees in the barn owl, compared to 340 degrees in the pigeon), which is compensated for the ability to turn the head horizontally by at least 270 degrees.
- To maintain a sharp focus when diving towards their prey, owls adjust the shape of the lens and cornea, but do so less efficiently than do humans. For some unknown reason, the Australian Tytonidae may have a poorer capacity to maintain a sharp focus than the American barn owl.

This brief description of visual competence indicates that selection for acuity was probably less intense than selection for hearing capacity to detect prey, as even a blind barn owl can forage in a familiar habitat. The neurophysiological visual system is indeed not highly complex, probably because nocturnal activity does not require a very detailed representation of the environment. Retinal ganglion cells are specialized neuronal cells that transmit the image from the retina to the optic tectum for visual processing. Barn owls have 4000 of these cells per square millimetre, compared with 10 000 per square millimetre in the diurnal burrowing owl. The visual and hearing systems apparently co-evolved. Owls do not move their eyes, which makes it easier for them to coordinate auditory and visual systems to determine prey location. By locking together the eyes and ears, the spatial reference frame is more stable, which facilitates prey location.

## Behavioural adaptations to nocturnal life
While it is amusing when owls bob and weave their heads in different directions, this is not to make us laugh. This behaviour compensates for the morphological and physiological limitations in visually locating prey or a predator and determining the distance and direction from which a sound has been emitted. When neither the prey nor the owl is moving, memory of the position of the prey fades because the prey is no longer providing auditory and visual information about its position. To actively generate a movement relative

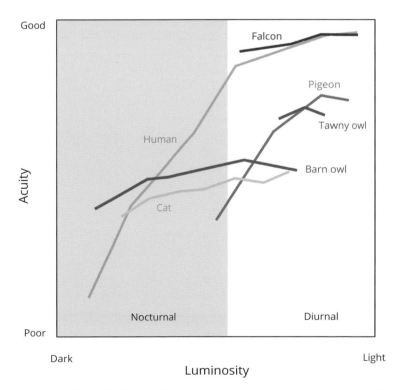

At night, barn owls have better acuity than humans and cats, whereas during the day, humans, falcons, pigeons and tawny owls outcompete barn owls. Barn owls have similar acuity under low (nocturnal vision) and high (diurnal vision) light. From Orlowski *et al.* 2012.

In the British Isles, barn owls are often seen flying during daylight hours. © Steve Nesbitt

to the retina and facilitate prey localization, owls move their heads with large horizontal and vertical movements, to take advantage of a phenomenon called **motion parallax**. When the prey suddenly moves, these large horizontal and vertical head movements allow the bird to scan a larger area, which decreases the likelihood of the prey leaving the owl's field of vision. When prey moves in any direction, the owl moves its head horizontally, often in a saccadic (jerky) motion, to follow the movement of the prey and to maintain its image in one position relative to the retina. Vertical head movements are related to the direction of prey movement: minimal vertical movement when the prey is moving sideways and significant vertical movement if the prey is moving towards or away from the owl.

## FUTURE RESEARCH
- The barn owl's visual capacity has been studied in great detail, yet little is known about inter-individual and inter-population differences in capacity to locate prey by sight or sound. There might be differences between individuals adopting alternative foraging strategies – with, for example, some individuals hunting more often by flying and others by using the sit-and-wait method. Barn owls from different populations may differ in their visual (and hearing) capacity because they prey upon different species requiring different hunting techniques.
- Nothing is known about the barn owl's ability to smell prey while flying slowly just above the ground.

## FURTHER READING
Fux, M. and Eilam, D. 2009. How barn owls (*Tyto alba*) visually follow moving voles (*Microtus socialis*) before attacking them. *Physiol. Behav.* **98**: 359–366.

Gutierrez-Ibanez, C., Iwaniuk, A. N., Lisney, T. J. and Wylie, D. R. 2013. Comparative study of visual pathways in owls (Aves: Strigiformes). *Brain Behav. Evol.* **81**: 27–39.

Harmening, W. M. and Wagner, H. 2011. From optic to attention: visual perception in barn owls. *J. Comp. Physiol. A* **197**: 1031–1042.

Harmening, W. M., Orlowski, J., Ben-Shahar, O. and Wagner, H. 2011. Overt attention towards oriented objects in free-viewing barn owls. *Proc. Natl. Acad. Sci. USA* **108**: 8461–8466.

Hausmann, L., Platchta, D. T. T., Singheiser, M., Brill, S. and Wagner, H. 2008. In-flight corrections in free-flying barn owls (*Tyto alba*) during sound localization tasks. *J. Exp. Biol.* **211**: 2976–2988.

Knudsen, E. I. 1982. Auditory and visual maps of space in the optic tectum of the owl. *J. Neurosci.* **2**: 1177–1194.

Land, M. F., Nilsson, D.-E. 2012. *Animal Eyes*. Oxford: Oxford University Press.

Orlowski, J., Harmening, W. and Wagner, H. 2012. Night vision in barn owls: visual acuity and contrast sensitivity under dark adaptation. *J. Vision* **12**: 1–8.

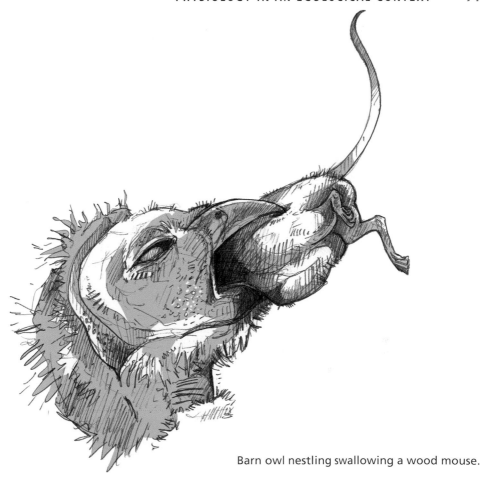

Barn owl nestling swallowing a wood mouse.

## 4.3   DAILY FOOD REQUIREMENTS

# A voracious predator

**How many mice does a barn owl eat in a year? This question is of interest not only to ecologists but to farmers as well. Barn owls consume three to four rodents per day; thus, a single family of barn owls can devour between 5000 and 7000 small mammals in a year. This makes the barn owl a highly valuable pest control agent.**

Calculating the number of rats, mice or voles consumed in a year is key to evaluating the ecological significance of the barn owl. This is not an easy exercise, because daily food consumption varies with environmental factors, such as precipitation and ambient temperatures, which is not surprising given that owls derive energy and water from their food. Food consumption increases in cold temperatures to counterbalance heat loss: barn owls can eat 80 grams of fresh meat per day at 25 °C and more (120 grams) at 5 °C. Food quality also strongly impacts daily food consumption: a captive owl relying on a diet of mice consumes twice as much as an owl that is fed on more energetically rich day-old poultry.

To estimate the number of rodents consumed in a year, let us consider the daily food consumption of French barn owls. At 20 °C, an adult consumes 75 grams of fresh small mammals per day, while a 20-day-old nestlings consume 64 grams per day, and up to 80 grams per day when it reaches

**6000 prey items** : total required to raise a family of four

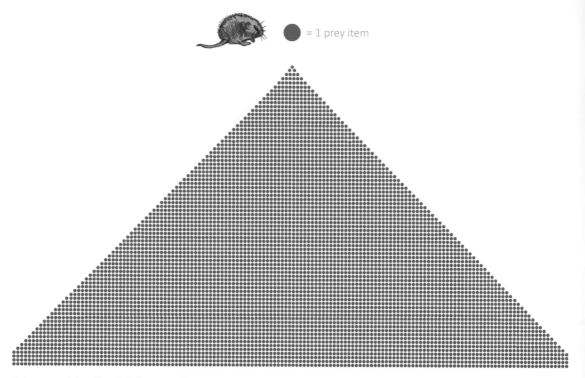

● = 1 prey item

A barn owl family with four nestlings can consume 6000 prey items in a year, as illustrated by this pyramid of 6000 dots.

60 days old. With a vole weighing 20–25 grams, each barn owl therefore consumes three to four voles daily. Therefore, a family with two to eight offspring devours 1200–3500 small mammals during the rearing period, and assuming that two of the offspring survive until the following year, the family could further consume 3000–5000 small mammals. Thus, each pair eliminates between 5000 and 7000 small mammals in one year. The barn owl is indeed the farmer's best friend!

## Poor digestive efficiency Compared to other owls and raptors, the barn owl appears voracious. Like many owls, the barn owl usually eats prey whole, including the bones and fur, from which no energy is derived. Tytonidae also show poor digestive efficiency, which may contribute to their large consumption of small mammals. An experiment in which owls and raptors were fed the same amount of mice demonstrated that fewer bones are fully digested by barn owls than by other birds: 825 bones were retrieved from barn owl pellets, 786 from great horned owl pellets, 748 from screech owl pellets, 176 from red-tailed hawk pellets, 36 from sparrowhawk pellets and 18 from rough-legged

buzzard pellets. The high rejection rate of bones indicates that the digestive juice found in the barn owl's stomach is only weakly acidic. This makes barn owl pellets very convenient tools for studying diet, given that prey species are identified by the bones retrieved from pellets.

The poor digestive ability is explained notably by the short small intestine of the barn owl compared with that of other owls and raptors, implying that the time available for nutrients to be absorbed by the body is shorter in the barn owl than in other birds. Accordingly, in the phylogenetically related masked owl, pellets contain 46% of the total sodium ingested, compared with 14% in the goshawk, swamp harrier and brown falcon.

Nestlings reach their final skeleton size a bit before they take their first flight. In order to accumulate calcium for bones, barn owls apparently change the efficiency of their digestion as they mature. This is probably why barn owl nestlings are better at digesting bones than adults, because of the higher acidity of their gastric juices. Interestingly, digestive efficiency is higher in nestlings aged 4–5 weeks than in older nestlings, which corresponds to a reduction in bone growth. This has been demonstrated by the fact that fewer bones are recovered in pellets ejected by young than old nestlings.

**Short intestine** The small intestine is responsible for food absorption, and the longer the intestine, the higher the digestive efficiency, which varies from 75% to 82% in birds of prey. The buzzard, a generalist predator, soars to forage, a technique that does not require speed. Hence, its body does not need to be exceptionally light, and its long intestine does not jeopardize foraging. Another species with relatively long intestines is the tawny owl, which hunts mainly by perching and dropping onto prey from above. This foraging technique is again not very energetically demanding, and therefore does not necessitate a low body mass for efficiency.

Short intestines evolved in two categories of raptors and owls. Hunting birds or other agile prey requires a fast upward stroke, which is not possible with a long intestine full of food. This is the case in sparrowhawks and peregrine falcons, which chase birds at high speed. The low digestive efficiency of these bird eaters is compensated for by the high nutritional value of avian prey. The second category includes birds, such as harriers and the barn owl, which hunt by flying slowly just above the ground. While an extremely short intestine helps reduce body mass, it also reduces digestive efficiency. This is one reason why barn owls produce pellets containing undigested bones of their prey.

## FUTURE RESEARCH
- The daily food consumption of free-living barn owls should be measured in both adults and nestlings. So far, all estimates come from birds kept in captivity, implying that the actual figures of daily food consumption may be blurred by stress induced by captivity, food quality and reduced physical activity.
- To further elucidate the ecological significance of the barn owl relative to that of other birds, food consumption, digestion duration and pH of the gastric juices of various owls and raptors should be measured under standardized conditions.
- Digestive efficiency should be measured in relation to parameters such as food type and the time of day or night when the meal is consumed.

## FURTHER READING
Barton, N. W. H. and Houston, D. C. 1993. A comparison of digestive efficiency in birds of prey. *Ibis* **135**: 363–371.

Barton, N. W. H. and Houston, D. C. 1994. Morphological adaptation of the digestive tract in relation to feeding ecology of raptors. *J. Zool.* **232**: 133–150.

Durant, J. M. and Handrich, Y. 1998. Growth and food requirement flexibility in captive chicks of the European barn owl (*Tyto alba*). *J. Zool.* **245**: 137–145.

Hamilton, K. L. 1985. Food and energy requirements of captive barn owls *Tyto alba*. *Comp. Biochem. Physiol. A* **80**: 355–358.

Hoffman, R. 1988. The contribution of raptorial birds to patterning in small mammal assemblages. *Paleobiology* **14**: 81–90.

Pellets are ejected orally and contain fur and bones of the owl's mammalian prey.

## 4.4  PELLET PRODUCTION

# A skeleton in the closet

**Pellets consist of the undigested body parts of prey, mostly bones surrounded by fur. Pellet production may have evolved to reduce the cost of digestion, to prevent the absorption of some nutrients, or to flexibly control variation in body mass. Owls are indeed able to finely regulate the time lag between two meals by accelerating or delaying pellet ejection.**

Why do owls, raptors, herons and gulls orally expel pellets containing remnants of previous meals? Is pellet production a by-product of the reduced length of the digestive tract that may have evolved to reduce body mass while flying? Or is it because the cloaca is the combined end point of the urinary, genital and digestive canals? Although answers to these questions have yet to be uncovered, the following four hypotheses based on adaptive arguments are worth considering:

- **Eliminating superfluous nutrients.** Some nutrients, such as calcium in the bones and keratin in the fur of prey, would become toxic if absorbed in large quantities. Oral elimination of these body parts would be more efficient than digestion and subsequent elimination through faeces. A corollary of this hypothesis is that owls often tear flesh from bones and fur to avoid eating these body parts, something that has not been studied in detail.
- **Cleaning the digestive tract.** The passage of pellets along the digestive canal during regurgitation helps to expel parasites and other pathogens, as well as undesirable items consumed by mistake.

- **Increasing digestive efficiency.** Once most of the necessary nutrients have been absorbed from a meal, further investment in digestion entails energetic costs not compensated for by the absorption of a few extra nutrients. Expelling undigested parts of a meal would effectively stop digestion.
- **Controlling variation in body mass.** A low body mass certainly helps when carrying a heavy prey item some distance to the nest. Moreover, actively reducing body mass is advantageous before resuming hunting, because any extra weight entails flying costs and reduces manoeuvrability. In this context, eliminating the unwanted elements from previous meals by expelling them inside a pellet is quicker and more efficient than digesting them fully and packing them in droppings. The need to rapidly reduce body mass may have selected for an ability to control the timing between food consumption and pellet expulsion.

While producing pellets confers some advantages, it poses at least three challenges. The first problem is the risk of suffocation, although a dead owl with a pellet stuck in its gape has been reported only twice, in France and Switzerland. A second problem is that passing the bones through the digestive canal risks damaging sensitive tissues. Another issue is the difficulty of eating the next meal before a pellet has been expelled. To solve this problem, barn owls can expel a pellet more rapidly (in 6.5 hours instead of 7.5–14 hours) when more food is imminently expected. Even sighting a vole can induce the ejection of a pellet in anticipation of a meal. And in captivity, owls can associate the presence of humans with the delivery of food, which induces them to eject a pellet.

In addition, barn owls can delay pellet production in order to consume several prey items in a row. This may explain why in the American barn owl the duration between food consumption and pellet ejection is relatively short (7.5–14 hours) if food is consumed in the morning or early afternoon, probably in anticipation of an evening meal. In contrast, if food is consumed in the late afternoon or at night, pellet production is delayed (11.5–16 hours), probably because it might be better to eat several prey items in a row at night and produce a pellet afterwards. For these reasons, the number of pellets produced per 24 hours can vary. Some authors mention one nocturnal and one diurnal pellet, but the situation may be more complex, given that the mean number of pellets produced per 24 hours varies from 1.2 to 2.7.

## FUTURE RESEARCH
- The number of pellets that nestlings produce per 24 hours should be determined in relation to variation in parental feeding rates.
- Is digestive efficiency related to nutrient requirement and body condition? For example, do owlets that are weakly competitive (e.g. the last-born individuals of a brood) or that experience poor feeding conditions take more time to produce a pellet, so as to absorb more nutrients?

## FURTHER READING
Barton, N. W. H and Houston, D. C. 1994. Morphological adaptation of the digestive tract in relation to feeding ecology of raptors. *J. Zool.* **232**: 133–150.

Juillard, M. 1979. Une cause particulière de mortalité juvénile chez la chouette effraie. *Nos Oiseaux* **35**: 37–39.

Marti, C. D. 1973. Food consumption and pellet formation rates in four owl species. *Wilson Bull.* **85**: 178–181.

Smith, C. R. and Richmond, M. E. 1972. Factors influencing pellet egestion and gastric pH in the barn owl. *Wilson Bull.* **84**: 179–186.

A thick layer of snow on the ground for a long period causes catastrophic reductions in barn owl population sizes.

## 4.5  CAPACITY TO WITHSTAND COLD WEATHER

# Not yet ready to conquer the Viking lands!

**Compared to other raptors, the barn owl is particularly sensitive to cold. The tropical origin of this bird suggests that it was not resistant to cold weather at the time that it was expanding through temperate regions, but why did the barn owl not evolve the capacity to withstand such climates? It may be that selection for high reproductive potential promotes the persistence of morphology, physiology and behaviour that are not compatible with resistance to cold. This situation may persist because high reproductive potential compensates for mortality caused by harsh winter weather.**

The barn owl is famous for its catastrophic reductions in population size after harsh winters. Populations can be totally decimated. When snow cover deeper than 5 centimetres persists for long periods of time, small mammals stay under this blanket. The difficulty in capturing these and alternative prey, such as small passerines, results in starvation – and consequently reduced owl survival and reduced population size. This explains why barn owls breed at low altitude and latitude, although there are exceptions, such as in France where a pair bred at 1500 metres in the Alpine foothills. On other continents, such as South America, barn owls can be found at higher altitudes where the weather is relatively clement.

Although, in early winter, French barn owls accumulate as much lipid (4–66 grams) as do long-eared and tawny owls, their survival prospects are still lower. If a lack of fat is not the cause of low winter survival, why are barn owls so susceptible to harsh winters? **Poor insulation due to a thin coat and no trousers!** Barn owls have fewer body feathers than do other species adapted to colder climates, such as tawny owls, and are one of the few owl species inhabiting temperate regions that do not have feathers on their legs, from where they lose substantial amounts of heat. The ambient temperatures at which barn owls do not need to invest extra energy to maintain a constant body temperature are

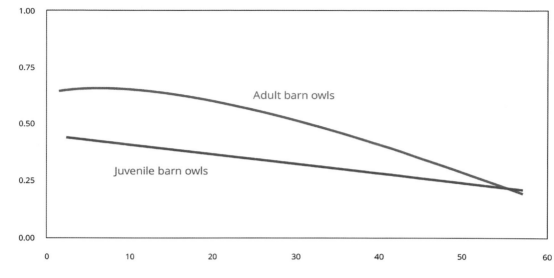

Survival of adult and juvenile barn owls in Switzerland in relation to the number of days with more than 5 centimetres of snow on the ground. From Altwegg *et al.* 2006.

high, ranging from 23 to 32 °C. For comparison, in the long-eared owl, the lowest temperature of this 'thermoneutral zone' is 17 °C, which further demonstrates the barn owl's high sensitivity to cold.

Therefore, **it is quite energetically costly for the barn owl to stay warm.** Compared to long-eared owls, barn owls show higher daily energy expenditure. Maintaining the basal metabolic rate during periods of food depletion and low ambient temperatures may become particularly expensive. A thick layer of fat insulates the body against cold, and although burning this layer of fat to warm up is effective, it is only a temporary solution to resist winter severity.

**Why not migrate to warmer regions in the winter?** In North America barn owls can migrate to reach more clement winter quarters, but this is surprisingly not the case in Europe.

## Starvation
The observed differences between barn owls and other raptors motivated French researchers to measure how barn owls can physiologically cope with fasting. When experimentally deprived of food, barn owls could fast for 3–15 days (the average was 9 days) at a temperature of 5 °C. At the end of the fasting period, owls were still able to fly, but their digestive capacity and locomotor activity were reduced. Amazingly, once re-fed, owls could restore their initial body mass within eight days, and only five weeks later they were laying as many eggs as they were the year before. This suggests that during years of food shortage or after a harsh winter owls lay few eggs relatively late in the season, not because breeding females are in poor body condition but to strategically adjust brood size to the prevailing environmental conditions.

Fasting can be divided into three distinct periods. Phase I lasts less than a day and is characterized by high body mass loss (20.5 grams per day). During the following seven-day phase II, body mass loss is less pronounced (8.6 grams per day) because lipids have been consumed and proteins are spared to preserve the ability to hunt. Close to death (phase III), there is an abrupt body mass loss (13.1 grams per day) during which the last protein reserves are consumed and lipids from the bone marrow mobilized. While fasting in the cold, barn owls reduce their resting metabolic rate by 16% during the first day, but maintain their body temperature at approximately 41 °C. Body temperature decreases only slightly (i.e. hypothermia, reaching a minimum of ca. 36 °C) close to death. Therefore, to resist the cold barn owls cannot reduce their basal metabolic rate and they must rely on expensive heat production (thermogenesis), which is costly and does not last long.

A barn owl sitting on a pole during a snowy day in the United Kingdom. © David Tipling

Although female barn owls are on average heavier than males, the chain of events occurring during fasting, as described above, is similar in the two sexes, causing a 30% body mass loss – corresponding to approximately 100 grams in a bird initially weighing 300–340 grams. The two sexes may therefore be equally resistant to starvation, potentially explaining why females were found dead as often as males during harsh winters in France, Denmark and the United Kingdom.

## Why is the barn owl not very well adapted to the cold? Tytonidae originate in the tropics, and from there they expanded to colder regions. Are barn owls in temperate regions sensitive to cold because they did not have time to evolve specific adaptations? Is the evolution of resistance to cold slowed down by immigrants coming from warm regions where they are not adapted to cold weather? Birds moving from southern Europe or America towards the northern parts of these continents bring genes adapted to southern climates, effectively preventing northern populations from evolving a higher capacity to resist harsh winters. This is plausible, because there are substantial exchanges of individuals between northern and southern populations on these two continents. However, it may not be sufficient to explain why the barn owl is so sensitive to harsh winters.

Tytonidae have a high reproductive potential, producing more than one brood per year, with each brood containing up to 9–11 offspring. The need for high foraging efficiency to feed large families may select for a low capacity to store fat because a lower body mass facilitates hunting. Birds that evolved the ability to reproduce at a high rate at the expense of cold resistance may have a selective advantage because producing many offspring should compensate for the reduction in survival during harsh winters.

## FUTURE RESEARCH
- The insulating properties of feathers from barn owls located in warm and cold climates should be investigated. This may provide information about whether barn owl feathers evolved specific adaptations to resist cold in temperate regions.

## FURTHER READING

Altwegg, R., Roulin, A., Kestenholz, M. and Jenni, L. 2006. Demographic effects of extreme winter weather in the barn owl. *Oecologia* **149**: 44–51.

Antoniazza, S., Burri, R., Fumagalli, L., Goudet, J. and Roulin, A. 2010. Local adaptation maintains clinal variation in melanin-based coloration of European barn owls (*Tyto alba*). *Evolution* **64**: 1944–1954.

Chausson, A., Henry, I., Ducret, B., Almasi, B. and Roulin, A. 2014. Tawny owl *Strix aluco* as a bioindicator of barn owl *Tyto alba* breeding and the effect of winter severity on barn owl reproduction. *Ibis* **156**: 433–441.

Handrich, Y., Nicolas, L. and Le Maho, Y. 1993. Winter starvation in captive common barn-owls: physiological states and reversible limits. *Auk* **110**: 458–469.

Handrich, Y., Nicolas, L. and Le Maho, Y. 1993. Winter starvation in captive common barn-owls: bioenergetics during refeeding. *Auk* **110**: 470–480.

Huang, A. C., Elliot, J. E., Cheng, K. M., Ritland, K., Ritland, C. E., Thomsen, S. K., Hindmarch, S. and Martin, A. 2015. Barn owl (*Tyto alba*) in western North America: phylogeographic structure, connectivity, and genetic diversity. *Conserv. Genet.* **17**: 357–367.

Massemin, S. and Handrich, Y. 1997. Higher winter mortality of the barn owl compared to the long-eared owl and the tawny owl: influence of lipid reserves and insulation? *Condor* **99**: 969–971.

Massemin, S., Groscolas, R. and Handrich, Y. 1997. Body composition of the European barn owl during the nonbreeding season. *Condor* **99**: 789–797.

Nielsen, M. L., Dichmann, K., Dahl, C., Lerche-Jorgensen, M., Settepani, V. and Erritzøe, J. 2014. Winter starvation in Danish barn owls. *Dan. Ornitol. Foren. Tidsskr.* **108**: 164–170.

Thouzeau, C., Duchamp, C. and Handrich, Y. 1999. Energy metabolism and body temperature of barn owls fasting in the cold. *Physiol. Biochem. Zool.* **72**: 170–178.

# 5 - Morphology

## in an ecological context

5.1  Body size
5.2  Reversed sexual size dimorphism

The large
Tasmanian masked
owl and a small
barn owl from
Galápagos.

## 5.1    BODY SIZE

# Gulliver and the Lilliputians

**Variation in body size is under strong genetic control; thus, a larger or smaller body size can quickly evolve if it confers new advantages under natural or sexual selection. Studying geographic variation in body size can provide useful information about the ecological factors that drive the evolution of a large or small body size.**

Barn owls vary in body size from small in the Galápagos Islands to very large in Tasmania. The smallest *Tyto* species on earth is on average half the size of the largest *Tyto* species. Why is there such a marked difference in size between different *Tyto* species?

To evaluate the potential role of ecological and climatic variables, the geographic distribution of differently sized species was compared. An analysis of 11 000 individuals preserved in 141 natural history museums all around the world showed that barn owl species are on average larger in wet regions than in dry regions, larger in the northern hemisphere than in the southern hemisphere, and regardless of hemisphere barn owls are larger towards the equator than towards temperate regions. Adding to these complex patterns, grass owls are larger in southern Africa than in northern Africa, while the opposite pattern is observed for the African barn owl. In South America, geographic variation in body size is extremely pronounced, with barn owls located in the southern parts of the continent being much smaller than those located closer to the equator. How natural selection generated these patterns of geographic variation in body size remains unknown. One thing is for sure: there is still room for research!

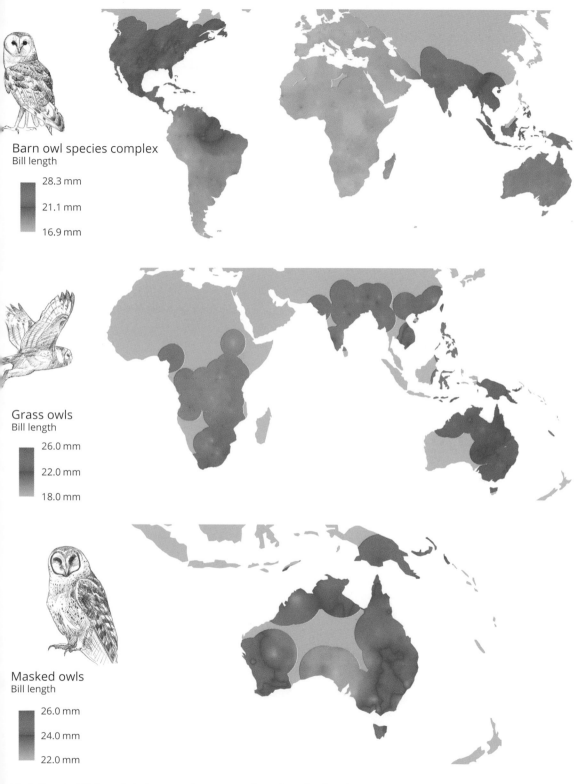

Barn owl species complex
Bill length

28.3 mm
21.1 mm
16.9 mm

Grass owls
Bill length

26.0 mm
22.0 mm
18.0 mm

Masked owls
Bill length

26.0 mm
24.0 mm
22.0 mm

Variation in bill length in barn owls, grass owls and masked owls.

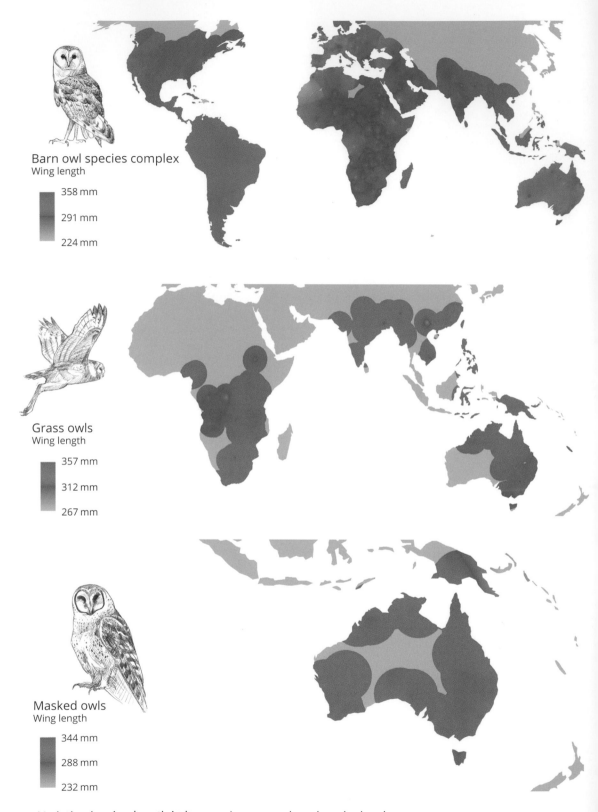

Barn owl species complex
Wing length

358 mm

291 mm

224 mm

Grass owls
Wing length

357 mm

312 mm

267 mm

Masked owls
Wing length

344 mm

288 mm

232 mm

Variation in wing length in barn owls, grass owls and masked owls.

Masked and sooty owls have much longer bills (a proxy of overall skeleton body size) than do barn owls and grass owls, but they do not necessarily have longer wings. African and European barn owls are particularly small compared to their counterparts on other continents. The same pattern is observed between African and Asiatic grass owls. These two graphs indicate that the largest Tytonidae in terms of bill size can display short wings (e.g. sooty owls) and that the smallest barn owls can have long wings (e.g. African and European barn owls). This is based on the measurement of skin specimens in natural history museums.

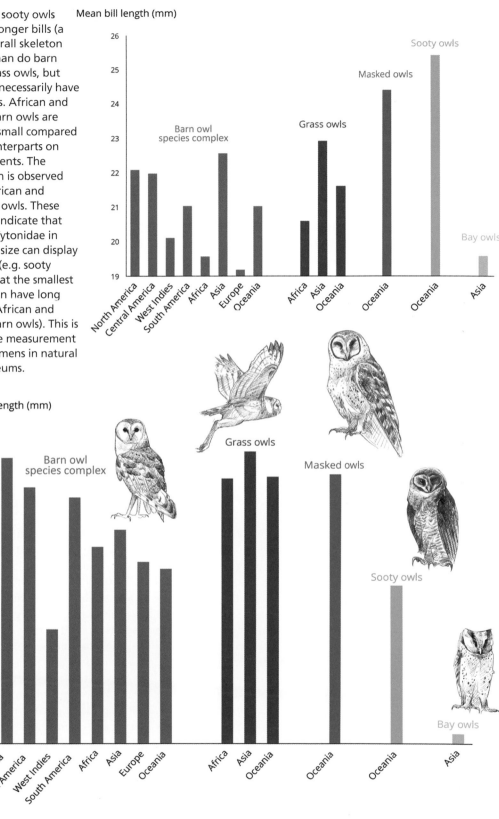

**Island syndrome** In the past, many Tytonidae lived on islands and were very large, up to 1.65 times larger than the current European barn owl and even larger than the current eagle owls. On islands, there is usually less food, fewer predators, and lower species diversity, and animals are less numerous than on the mainland. These ecological conditions select for different sizes than on the mainland in many organisms, and the barn owl is no exception. Owls on remote islands have a larger skeleton than owls located on islands close to a continent, but not necessarily longer wings or tails. Was a large body size a prerequisite for flying to remote islands?

Once the barn owl reached an island, selection promoted the evolution of shorter wings and tails, regardless of the distance to the nearest continent. Opportunities to explore new regions are more limited on islands than on continents; thus, longer wings and tails may not be necessary on islands. This hypothesis is supported by owls located on large islands having longer wings, tails and bills than those on smaller islands, where there is less space to fly and explore.

**Environmental and genetic determinism of body size** If tall human parents have tall children, is this pattern also the same in the barn owl? A strong genetic basis of body size implies that barn owls could quickly evolve towards a larger (or smaller) body size if a large (or small) size suddenly provides reproductive or survival advantages.

A classic approach to discovering whether the resemblance in body size between parents and offspring is genetic or environmentally determined is to experimentally swap eggs or hatchlings between randomly chosen nests. If the cross-fostered offspring reach a similar size to their biological parents, we can conclude that they inherited genes regulating body size. If the cross-fostered offspring reach a similar size to their foster parents, we can assume that growth is sensitive to the rearing environment and that the rearing conditions experienced by the cross-fostered nestlings are like those experienced by their foster parents when they were young. Even if offspring do not develop to a size like that of their foster parents, the rearing environment may still influence body size. This can be detected if unrelated offspring sharing the same rearing environment develop a similar size because they receive a similar amount of food or food of similar quality.

Cross-fostering experiments have been repeatedly performed in Switzerland. These studies showed that the bills of offspring and biological parents are of similar length (heritability estimate of bill length, $h^2 = 0.62 \pm 0.19$), a conclusion that also applies to claw length ($h^2 = 0.62 \pm 0.18$), wing length ($h^2 = 0.48 \pm 0.17$) and tail length ($h^2 = 0.42 \pm 0.19$) but to a lesser extent to tarsus length ($h^2 = 0.25 \pm 0.24$).

Genetics thus has a strong influence on body size, but what about the rearing environment? Two pieces of evidence confirm that the rearing environment and hunting ability of the adults that rear the nestlings has a pronounced impact on body size. First, unrelated individuals raised in the same nest develop to a similar size. Second, the first-born nestlings are better able to monopolize food than their later-born siblings and thus reach a larger body size. However, there is no evidence that foster offspring resemble their foster parents with respect to body size, indicating that larger parents do not provide better parental care than smaller parents. If larger parents brought more food or food of higher quality than smaller parents, foster offspring should have grown to a size like that of their foster parents, which was not the case in Switzerland.

**Abnormal morphologies** Morphological aberrations are rarely reported in the literature, because abnormal individuals usually die before they reach breeding age. Rare phenotypes can be the outcome of stressful environmental conditions experienced during growth, or they may be due to genetic mutations. Genetic variation for rare phenotypes can be maintained if these phenotypes are not too detrimental, as shown by human vestigial structures that lost their function through evolution, such as the coccyx or the appendix.

Very few cases of abnormal morphology in the barn owl have been reported in the literature. These cases include an individual with five instead of four toes on each leg, one born with a single wing, and one blind nestling, and although 97% of barns owls in Switzerland have twelve tail feathers, they can sometimes have ten, eleven, thirteen or fourteen feathers. The underlying genetic variation that produces aberrant individuals can provide the raw material to evolve new phenotypes. Perhaps at some point during evolution having more or fewer than twelve tail feathers may have provided selective advantages, thus explaining the maintenance of this phenotype in the population.

## FUTURE RESEARCH

- In terms of body size, is natural selection directional (e.g. do large individuals have a higher fitness than smaller conspecifics) or stabilizing (are individuals of medium size more fit than small or large individuals)? Or does natural selection fluctuate over time depending on ecological or climatic conditions? For instance, a large body size may be beneficial in warm years and a small size in cold years.
- Why are masked and sooty owls so large? More analyses should be performed to identify the ecological and climatic factors that account for geographic variation in body size.

## FURTHER READING

Dürr, T. 1993. Abnormalities in the number of tail feathers in the barn owl *Tyto alba* and the willow warbler *Phylloscopus trochilus*. *Beitr. Vogelkd* **1**: 60–64.

Grant, P. R. 1998. *Evolution on Islands*. Oxford: Oxford University Press.

Kumerloeve, H. 1968. Eine beiderseits fünfzehige Schleiereule, *Tyto alba guttata* (Chr. L. Brehm): ein seltener Fall von Hyperdaktylie. *Bonn. Zoolog. Beitr.* **19**: 211–214.

Roulin, A. 2006. Linkage disequilibrium between a melanin-based colour polymorphism and tail length in the barn owl. *Biol. J. Linn. Soc.* **88**: 475–488.

Roulin, A. and Salamin, N. 2010. Insularity and the evolution of melanism, sexual dichromatism and body size in the worldwide-distributed barn owl. *J. Evol. Biol.* **23**: 925–934.

Roulin, A., Wink, M. and Salamin, N. 2009. Selection on a eumelanic ornament is stronger in the tropics than in temperate zones in the worldwide-distributed barn owl. *J. Evol. Biol.* **22**: 345–354.

Scherner, E. 1974. Eine einflügelig geborene Schleiereule (*Tyto alba guttata*). *Beitr. Vogelkd* **20**: 451–456.

Schwarz, H. 1987. Fund einer blinden Schleiereule (*Tyto alba guttata*). *Beitr. Naturkd. Wetterau* **7**: 224–226.

The sooty owl shows the most pronounced reversed sexual size dimorphism in owls, with females weighing 1000–1200 grams and males weighing 500–700 grams.

## 5.2   REVERSED SEXUAL SIZE DIMORPHISM

# Big mom and skinny dad

Reversed sexual size dimorphism (RSD) describes species in which females are on average larger than males. This is an unusual situation in birds except for some families, including owls, hawks, falcons, skuas, boobies, frigatebirds and the polyandrous jacanas and phalaropes. Female owls are large and hence often capture larger prey than do the smaller males. RSD may thus allow a pair of owls to prey upon a wider range of animals.

In some Accipitridae (hawks), Falconidae (falcons) and Jacanidae (jacanas), females can weigh up to twice as much as males. In the extinct New Zealand moa, females were even up to 280% larger than males. Three non-mutually exclusive selective processes can explain why female owls and raptors are larger than males.

- **Sexual dimorphism hypothesis.** Males are selected to be small and females to be large. Males specialize in small prey and females in large prey, thereby reducing foraging competition between the two sexes. By hunting different prey, parents can exploit a larger ecological niche than if males and females were the same size. RSD would therefore allow parents to feed their chicks more efficiently.

- **Small-male hypothesis.** Selection to be small in males is stronger than selection to be large in females. The offspring need a lot of food during a relatively short period of time, and hence relying on large prey may not be optimal because large animals are not abundant and are heavy to transport to the nest. It might be more efficient to hunt small animals, which are much more abundant and easy to bring back to the nest. In this context, a small body size may confer an advantage, making it easier to capture agile prey items and reducing the cost of flying. Because the male provides most of the food to the offspring, a small body would allow him to sustain intense foraging activities. This hypothesis is supported by observations of Tengmalm's owls.
- **Large-female hypothesis.** Selection to be large is stronger in females than selection to be small in males. A large body size may facilitate female-specific reproductive activities, including the production of large eggs, incubation duties and nest protection while the male is foraging.

Limited data are available to discriminate between these three hypotheses, not only in the barn owl but in most bird species. However, pronounced RSD evolved in owl species (Strigidae and Tytonidae) catching both agile and large prey. It seems, therefore, that this pattern of sexual dimorphism did not evolve to facilitate female reproductive activities but rather to reduce foraging competition within the family: males capture small agile prey, and females catch larger prey. This interpretation

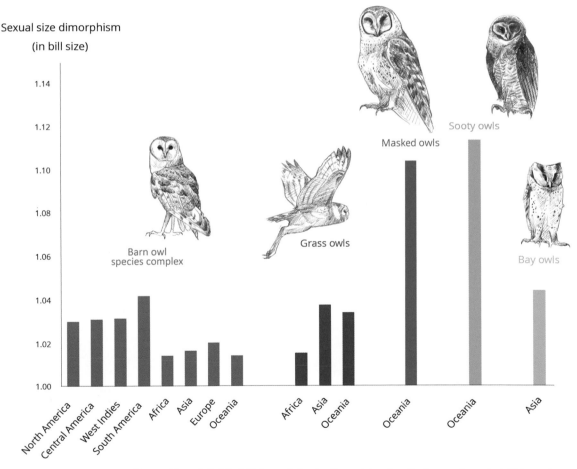

Reversed sexual size dimorphism (RSD) in bill length in the Tytonidae family (mean female bill length divided by mean male bill length). The sooty owls and masked owls are more sexually dimorphic than the other groups. This is based on the measurement of skin specimens in natural history museums.

is consistent with the sexual dimorphism hypothesis and results from a comparison between the degree of sexual size dimorphism and diet across owls.

What about the specific case of Tytonidae? Why are female sooty owls and Tasmanian masked owls much larger than males (body mass is 1000–1200 vs. 500–700 grams, and 880 vs. 580 grams, respectively), whereas female barn owls and African grass owls are only slightly larger than males (by about 5%)? This pattern is probably related to foraging behaviour, since in sooty owls the diets of males and females are very different, with an overlap of only 62%. Female sooty owls capture prey that is three to five times heavier than the prey of their male mates.

The size difference between males and females is not present at birth. In Switzerland, the body mass difference between male and female barn owl nestlings is less than 1% at 10 days of age, increasing to the adult value of 5% by 50 days.

## FUTURE RESEARCH
- The ecological and life-history correlates of variation in the degree of reversed sexual size dimorphism among Tytonidae should be identified.
- Whether size differences between male and female partners are correlated with reproductive parameters should be determined.
- Whether selection on body size is positive, negative or stabilizing should be tested.

## FURTHER READING
Almasi, B. and Roulin, A. 2015. Signalling value of maternal and paternal melanism in the barn owl: implication for the resolution of the lek paradox. *Biol. J. Linn. Soc.* **115**: 376–390.

Andersson, M. and Norberg, R. Å. 1981. Evolution of reversed sexual size dimorphism and role partitioning among predatory birds, with a size scaling of flight performance. *Biol. J. Linn. Soc.* **15**: 105–130.

Bilney, R. J. 2011. Potential competition between two top-order predators following a dramatic contraction in the diversity of their prey base. *Anim. Biol.* **61**: 29–47.

Hakkarainen, H. and Korpimäki, E. 1991. Reversed sexual size dimorphism in Tengmalm's owl: is small male size adaptive? *Oikos* **61**: 337–346.

Krüger, O. 2005. The evolution of reversed sexual size dimorphism in hawks, falcons and owls: a comparative study. *Evol. Ecol.* **19**: 467–486.

Mueller, H. C. 1986. The evolution of reversed sexual dimorphism in owls: an empirical analysis of possible selective factors. *Wilson Bull.* **98**: 387–406.

Paton, P. W. C., Messina, F. J. and Griffin, C. R. 1994. A phylogenetic approach to reversed size dimorphism in diurnal raptors. *Oikos* **71**: 492–498.

# 6 Daily life:

## hunting, feeding and sleeping

6.1   A nocturnal hunter
6.2   Roosting
6.3   Home range
6.4   Flight mechanics
6.5   Hunting methods
6.6   Prey selection
6.7   Diet

A barn owl inside a building.

## 6.1   A NOCTURNAL HUNTER

# Night owl

**Although owls are nocturnal par excellence, the situation in real life is slightly more complex. Throughout the world, almost all barn owls hunt only at night, but there are notable exceptions in Tonga, Samoa and the United Kingdom. Recording sleep patterns using non-invasive methods has shown that, although this bird is clearly nocturnal, the pattern is not so clear-cut, with adults often being awake during the day.**

For animals, the struggle to survive is tough. Predators are numerous and threatening, as they have evolved a battery of adaptations to be efficient hunters. Millions of years ago, in response to intense predation, but also to compete with the ectothermic dinosaurs that required sunlight to warm up, diurnal mammals evolved the ability to be active at night. Some dinosaurs also developed endothermy and evolved into what we now know as birds. Natural selection then favoured a temporal shift in activity between owls and other predatory birds to reduce foraging competition. For example, the diet of the barn owl is more like that of the eared owls (*Asio* spp.) (diet overlap is 0.60; a value of 1 would indicate that they capture the same prey species at the same frequency) than that of the diurnal falcons (0.48) and little owls (*Athene*) (0.36), which, although exploiting similar habitats, prey more often on lizards and invertebrates.

**Nocturnal versus diurnal**  Barn owls and their relatives are mainly nocturnal. They have occasionally been observed foraging during the day in South Africa, Argentina, Surinam and the USA, but the only places where they are regularly observed flying at all times of day and night is in Samoa and Tonga (but not Fiji) and in parts of the United Kingdom (but not in Ireland). If diurnality on Pacific islands may be due to the absence of competition with other raptors, why are barn owls often observed foraging before sunset and after sunrise in Great Britain but not elsewhere in Europe? Is this because they prey on field voles (*Microtus agrestis*) in Great Britain but on common voles (*M. arvalis*) on the mainland? Perhaps field voles are more active than common voles during daylight hours. Alternatively, is this because there is no full darkness in midsummer in the northern parts of the United Kingdom? There are more questions than there are answers.

Hunting in daylight may occur more frequently when food is scarce, particularly during unfavourable winter weather, or when foraging only at night is not sufficient to meet the food needs of the brood, or when the nights are not long enough. Unfortunately, no quantitative data are available to evaluate these hypotheses. What we know so far is that some individuals are active during the day more frequently than others, and that a given individual can fly during the daytime one day but not the next. In winter, out of nine English barn owls, eight were regular diurnal hunters, emerging on average 79 minutes before sunset and continuing to fly for up to 139 minutes after sunrise. A breeding male foraged during daylight on 24 out of 60 days (40%), and his female partner did so on 37 out of 69 days (54%). This behaviour is interesting and deserves additional research.

**Empty belly**  Having a stomach full of food may not be optimal if the goal is to fly slowly and stealthily. To circumvent this problem, male barn owls bring all prey items necessary to meet the daily food requirements of the entire family at the beginning of the night and then eat at the end of the night. Although eating after all other family members may seem to be an act of pure devotion to the family, this behaviour reduces the flying costs associated with foraging and hence enhances hunting success. Prey items delivered more rapidly than the offspring can eat them accumulate in the nest, allowing the nestlings to spread their meals over 24 hours, eating both at night and during the day. Thus, nestlings can organize the timing of their meals.

Accumulating prey in one place before eating it has never been observed outside the breeding season, probably because the risk that other individuals or animals will steal the stored food can be high. If creating a food depot is not an option, barn owls may finely adjust the time when they resume hunting in relation to the digestion of their last meal. Among dead barn owls found along highways in France, larger birds had more food in their stomachs than smaller ones, which suggests that only the largest individuals can afford to fly with the additional weight. These birds had up to 50 grams in their stomachs, which represents 16% of their body mass.

**Sleeping time**  Captive American barn owls are active mainly when the ambient light has an intensity of 0.2–0.4 lux, corresponding to full-moon nights, in contrast to the range of 4.6–155 lux observed for the diurnally and nocturnally active little owl. Probably because owls fast during daylight, their peak of activity occurs at the beginning of the night, when they set out to find food. The reasons for a second peak in activity just before sunrise, and for captive adults being relatively active around midday, remain unclear. These results, together with the observation that 15% of their flight occurs in daylight, indicate that barn owls are often awake during the day.

Daytime wakefulness was formally demonstrated in another set of adult owls by monitoring brain activity with electroencephalography, using a minimally invasive method that does not disturb the birds. This approach showed that, although adult barn owls are nocturnal, they are nevertheless cerebrally awake 29.5% of the time during the day and asleep at night 13.6% of the time. However, daytime seems to be the main sleeping period, as sleeping bouts were longer during the day than at night, while wakefulness bouts were longer at night than during the day. This indicates that despite some diurnal activity, barn owls mainly sleep during the day.

Sleep is quite fragmented in barn owls, as in other birds. While humans typically show four or five episodes of deep sleep – so-called rapid-eye-movement (REM) sleep – at night, an average of 370 REM episodes were detected during a 24-hour period in each of four measured barn owl adults. The duration of a single wakefulness bout was usually only 12 minutes, with the shortest bout being 4 seconds and the longest 1 hour, whereas the duration of sleep bouts was on average 1 minute, with the shortest being 4 seconds and the longest 27 minutes. It therefore seems that captive barn owls can accomplish what they need to during periods of only 12 minutes and recover with 1-minute-long sleeping episodes. The function of such a sleeping architecture is unclear, and the barn owl is surely an interesting model organism for studying sleep.

## FUTURE RESEARCH
- Whether barn owls are more likely to forage during the day in northern than in southern Great Britain should be investigated, in addition to whether the propensity for hunting during the day is related to winter feeding conditions, food need and reproductive activities.
- The timing of foraging during and outside of the breeding season in relation to factors such as light intensity, wind and food abundance should be monitored.
- The sleep activities of free-living adult barn owls should be recorded.

## FURTHER READING
Bunn, D. S., Warburton, T. and Wilson, R. D. S. 1982. *The Barn Owl*. Calton: T. & A. D. Poyser.
Charter, M., Leshem, Y., Izhaki, I. and Roulin, A. 2015. Pheomelanin-based colouration is correlated with indices of flying strategies in the barn owl. *J. Ornithol.* **156**: 309–312.
Durant, J. M., Hjermann, D. Ø. and Handrich, Y. 2013. Diel feeding strategy during breeding in male barn owls (*Tyto alba*). *J. Ornithol.* **154**: 863–869.
Erkert, H. G. 1969. Die Bedeutung des Lichtsinnes für Aktivität und Raumorientierung der Schleiereule (*Tyto alba guttata* Brehm). *Z. Vgl. Physiol.* **64**: 37–70.
Gerkema, M. P., Davies, W. I. L., Foster, R. G., Menaker, M. and Hut, R. A. 2013. The nocturnal bottleneck and the evolution of activity patterns in mammals. *Proc. R. Soc. Lond. B* **280**: 20130508.
Halle, S. and Lehmann, U. 1992. Cycle-correlated changes in the activity behavior of field voles, *Microtus agrestis*. *Oikos* **64**: 489–497.
Jaksic, F. M. 1982. Inadequacy of activity time as a niche difference: the case of diurnal and nocturnal raptors. *Oecologia* **52**: 171–175.
Roulin, A. 2004. The function of food stores in bird nests: observations and experiments in the barn owl *Tyto alba*. *Ardea* **92**: 69–78.
Scriba, M. F., Harmening, W. M., Mettke-Hofmann, C., Vyssotski, A. L., Roulin, A., Wagner, H. and Rattenborg, N. C. 2013. Evaluation of two minimally invasive techniques for electroencephalogram recording in wild or free behaving animals. *J. Comp. Physiol. A* **199**: 183–189.

A barn owl nestling sleeping.

## 6.2  ROOSTING

# Taking a nap

**Birds roost in safe places that provide shelter from adverse weather conditions and predators. Although most Tytonidae use a variety of sites including trees (on branches and/or in hollows), caves and man-made structures, grass owls sleep exclusively on the ground, sometimes in groups of up to twelve individuals.**

Owls roost in dedicated places out of reach of predators and the small birds that harass them. These safe places also protect against wind and rain and provide insulation from cold or hot weather. Each species of Tytonidae has a different habitat preference, and the selected roost sites vary across landscapes depending on availability. While sooty owls roost exclusively in trees and caves and grass owls on the ground among grasses, barn owls of the *Tyto alba* species complex (the western *T. alba*, the American *T. furcata* and the eastern *T. javanica*) are more eclectic in their roost choice.

**Daylight roost sites** Most barn owls of the *Tyto alba* species complex roost in man-made constructions, although roosting in trees can be frequent in some regions, such as in the United Kingdom. As observed in Germany, Canada and India, 70–90% of the roosts are in farms, temples and other buildings, usually between 4 and 27 metres above the ground. In contrast to roosting in trees, roosting inside buildings can conserve substantial energy, particularly on wet and windy days. This benefit may explain why barn owls only occasionally roost in trees. In France and Switzerland,

A barn owl roosting in a palm tree.

roosts inside forests have been observed on rare occasions at up to 600 metres from the forest edge. Around the globe, barn owl populations can also be seen roosting in caves, crevices and lava tubes, on the ground, or even underground, particularly in places devoid of trees and houses, such as on the Alegranza archipelago in the Canaries. Rocky sites are also preferentially used for roosts in Curaçao and Cape Verde. Unconventional roosting sites, such as reed beds, are also sometimes observed.

Detailed descriptive data on roost use are rare, except for the large sooty and masked owls. These data are important, because these big owls are threatened by human activities, and any insight into their daily lives may become important for their conservation. Sooty owls preferentially roost in dense rainforests in dark and sheltered places. Females rarely roost in foliage and prefer tree hollows and caves, whereas in males the opposite is observed. This difference between sexes may result from females being larger and more conspicuous when roosting in exposed locations. In contrast, males hunt at some distance from the nest and thus may need to be more flexible in roost-site selection. Cave roosts are observed slightly more often in cliffs than in gorges, with a preference for cliffs oriented away from the prevailing sunlight (darker and shadier sites). These cliffs may provide better climatic conditions and greater concealment from disturbance.

Although the Australian masked owl usually roosts alone in thick foliage in dense gullies and in hollows of eucalypts, it can also sleep in caves and crevices in cliffs, but apparently never in buildings. Roosts are located at a height of 3–20 metres above ground, and one bird was observed to use the same roost for 24 days in a row. Climatic conditions likely influence roost-site selection, which may explain why roosting behaviour is different in Tasmania compared to mainland Australia. In a Tasmanian study, masked owls were indeed found to roost in trees half of the time, usually on a branch covered with dense foliage and rarely inside a hollow, while they occasionally selected buildings (17%) and cliffs (37%). The Sulawesi masked owl (*Tyto rosenbergii*) is also eclectic in its roost

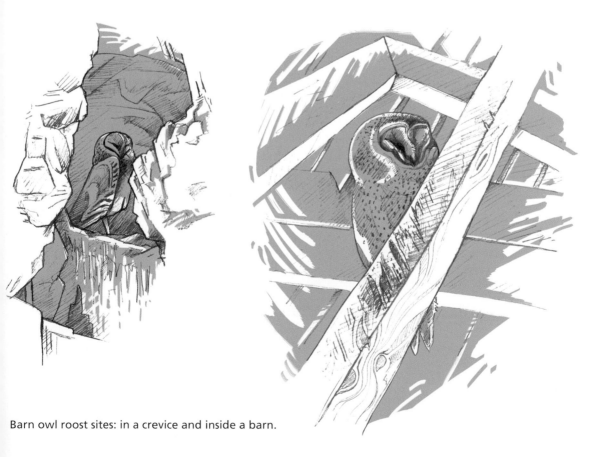

Barn owl roost sites: in a crevice and inside a barn.

A barn owl roosting
in a tree hole.

choice; it rests in lightly wooded cultivation, on tall dead trees, on coconut plantations, in caves and occasionally around villages.

**Frequency of use** Some barn owls habitually roost in the exact same place for extended periods of time, as observed in Germany, where four males used the same roost for up to 78 days in a row. In the barn owl species complex, which usually roost in man-made sites, changing roost sites is probably triggered by human disturbance. Indeed, although these owls generally prefer one roost, they often rotate between multiple sites. In Germany, a non-breeding female slept in six different locations over a period of 128 days, and one male used three roosts over 16 days. The observations of Australian sooty and masked owls are fundamentally different, as these owls regularly change roost sites to avoid predators or harassment. Radio-tracked sooty owls used between six and thirteen roosts over several months. Even though it is rare to observe these owls returning to the same place on consecutive nights, some roost sites are used by the same individual for several years. Suitable cave sites can even be used for thousands of years by different generations of owls.

**Distance from nest** Males usually roost further from the nest than females, because females must be close to their offspring to brood and to feed them, whereas males collect food far from the nest. Once the nestlings are old enough to maintain their own body heat and to feed without maternal help, the mother more frequently roosts outside the nest, usually within 1 kilometre.

As long as the adult barn owls are caring for their nestlings, they roost either with their young inside the nest cavity or up to 3–8 kilometres away from the nest, as observed in Europe and the Americas. In a German study, 41% of the roosts were located less than 500 metres from the nest, and not necessarily close to rich foraging sites – indicating that the properties of the roost site itself are more important than the quality of the habitat surrounding it. Thus, the distance between roost and nest will mainly depend on the proximity of suitable roost sites. The observations of the African and eastern grass owls are different because these owls roost on the ground in homogeneous open grasslands and can therefore find suitable roost sites quite close to their nests.

**Social roosts**  Usually, Tytonidae roost solitarily within their territory to defend their nesting cavity, sometimes year-round. In the barn owl, the only reported social roost was a pair of Scottish birds that were observed together during the breeding season with first-year unrelated visitors. Communal roosts have also been observed in Australia. In contrast, long-eared owls roost communally in groups of tens of conspecifics and sometimes up to one hundred, perhaps because they hunt in nearby fields where voles are particularly abundant and where the competition for food is low. Long-eared owls can be gregarious because they breed in old crow nests, which are abundant and hence do not need to be defended in winter. Similarly, grass owls are more social, with pairs of birds sometimes roosting together at any time of the year, and even close to marsh owls, as observed in South Africa. Small groups of up to five grass owls have been observed roosting together in Africa, and up to twelve individuals in Australia. As with the long-eared owls, this is most likely because nest-site availability is not a limiting factor for grass owls, and hence nest sites do not need to be defended year-round.

## FUTURE RESEARCH
- Study the frequency of use of various roost types across populations. This is necessary to inform conservation practice.
- Investigate whether some roosts are used mainly because they confer protection against predators or because they provide shelter from inclement weather conditions.
- Identify seasonal differences in roost-site selection.
- Do reddish and whitish owls use different roost sites? This might be so if plumage colouration plays a role in camouflage.

## FURTHER READING
Bilney, R. J., Cooke, R. and White, J. G. 2010. Underestimated and severe: small mammal decline from the forests of south-eastern Australia since European settlement, as revealed by a top-order predator. *Biol. Conserv.* **143**: 52–59.

Bilney, R. J., Cooke, R. and White, J. G. 2011. Potential competition between two top-order predators following a dramatic contraction in the diversity of their prey base. *Anim. Biol.* **61**: 29–47.

Brandt, T. and Seebass, C. 1994. *Die Schleiereule: Ökologie eines heimlichen Kulturfolgers*. Wiesbaden: Aula Verlag.

Fitzsimons, J. A. 2010. Notes on the roost sites of the Sulawesi masked owl *Tyto rosenbergii*. *Forktail* **26**: 142–145.

McCafferty, D. J., Moncrieff, J. B. and Taylor, I. R. 2001. How much energy do barn owls (*Tyto alba*) save by roosting. *J. Therm. Biol.* **26**: 193–203.

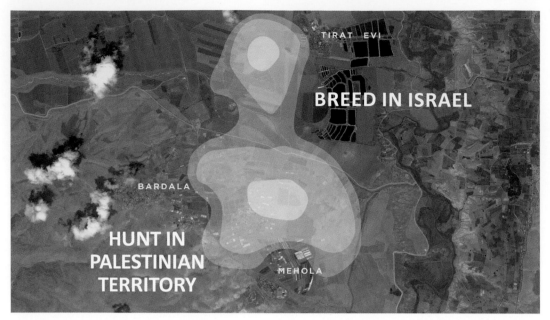

Home range of a pair of barn owls breeding in Israel and foraging mainly in the Palestinian territory. The whiter parts represent the regions where the owls spent most of the time. © Motti Charter

## 6.3  HOME RANGE

# Home sweet home

**Top predators typically exploit resources over large areas. However, their home range can be restricted to a few hundred metres around the nest when food is plentiful, even though barn owls can fly long distances when rodents or other prey are scarce in the nesting home range. Owls usually visit specific habitats where prey are detectable and easily accessible. While most barn owls exploit agricultural fields, several *Tyto* species hunt and breed in grasslands or in forests.**

The **home range** defines the area where an animal forages and roosts on a periodic basis. Knowing the size of the home range is useful for determining the number of individuals that could settle in a given region. When resources are scarce or when human disturbance is common, animals expand their home range to reproduce and survive. In contrast, a **territory** is the area defended by breeding individuals. Depending on social behaviour, a bird's territory may be identical to its home range (as in tawny owls), limited to the surroundings of its nesting site (as in little owls) or restricted to the nesting spot itself (as in colonial birds). Thanks to radio- and GPS-tracking devices, knowledge about animal movements is rapidly accumulating. While a home range can be easily quantified by registering the spatial behaviour of owls, quantifying territory size requires an assessment of territorial display, something that has yet to be done in Tytonidae.

**Home range size**  Breeding males can forage for their offspring in a territory as small as 0.9 square kilometres for Malaysian barn owls to up to 35 square kilometres for greater sooty owls. Barn owls can sometimes hunt quite far from their nest: 4.2 kilometres in Germany, 4.5 kilometres in Scotland, 5.6 kilometres in the USA – and 7.4 kilometres in the case of the greater sooty owl. In Switzerland, males cover on average 5 kilometres per hour when providing for their young, and in one night they can travel between 5 and 53 kilometres. When hunting they fly at a speed of about 18 kilometres per hour (5 m/s), and once they have captured a prey item they bring it back to the nest at a speed of 25 kilometres per hour (7 m/s).

As long as barn owls find food without too much difficulty, they can tolerate conspecifics inside their home range, and the home ranges of male and female breeding partners can overlap, as observed in France, Scotland and Switzerland. In the breeding season, females have smaller home ranges than males, because males must patrol larger areas to find food that females then distribute among the nestlings. In the greater sooty owl, the home ranges of male and female partners were 30.25 and 9.94 square kilometres, respectively.

Obviously, the ultimate goal for any animal is to raise a large high-quality family without the cost of travelling long distances to find food. Thus, only pairs that have access to a large quantity of food near their nest should be able to produce a large family. In Switzerland, pairs producing large clutches and broods flew over smaller home ranges than did less successful pairs, and furthermore, barn owl males in Scotland captured their prey closer to their nest in good vole years than in poor vole years.

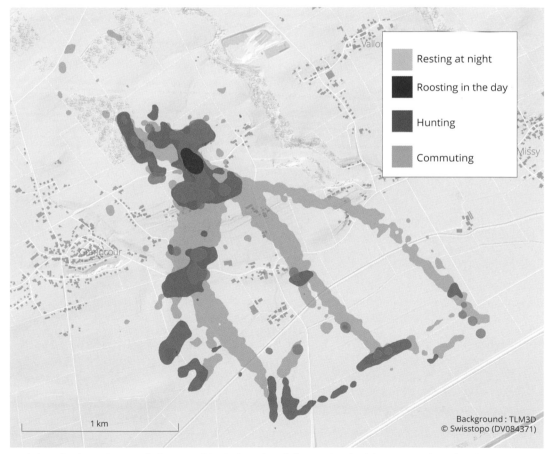

Activity of a breeding male barn owl in Switzerland determined with GPS. During the day, this bird roosted mainly close to its nest (in dark blue), while at night it rested in different places (in light blue). Once a prey item had been captured (foraging places in orange), the barn owl mainly used one of three paths to return to its nest (in grey, 'commuting').

Three habitats suitable for the barn owl: open landscapes (*left*), open forests (*middle*) and wet habitats (*right*).

Barn owls are often loyal to their home range over several years, breeding regularly at the same site. Do owls exploit their territory differently during and outside of the breeding season? When owls are free from parental duties, they can enlarge their home ranges as the offspring are no longer at home, implying that the ties that adults have with their nest site become loose. Furthermore, outside of the breeding season, prey are less abundant, which causes owls to forage over much larger areas. In Scotland, 90% of observations were within 1 kilometre of the nest in summertime, compared with only 40% in wintertime, and in Germany a female flew on average only 0.637 kilometres (maximum 3.500 kilometres) from her nest during the breeding season and then 2.164 kilometres (maximum 4.175 kilometres) after, increasing her home range from 4.62 to 6.93 square kilometres between June and November. Similar findings apply to the greater sooty owl, with the size of the home range being 1.5–2.2 times smaller in breeding than in non-breeding females.

## Foraging habitat

Identifying the exact habitats in which owls hunt is not easy, given their nocturnal habits. A few owls have been radio-tracked to provide useful information about where they hunt. Barn owls forage selectively in places where their prey is easily detectable and accessible, such as in fields with short grass. In arid environments, owls also avoid the more dangerous spiny bushes and prefer open areas for hunting.

The most preferred habitats in radio-tracked barn owls in Switzerland were, in order of preference: fields with cereals, grasslands, forest edges, settlements, riverside areas, fields of tobacco, maize, other crops, and wildflower areas. Similar tracking methods used in New Jersey and Ohio demonstrated that grasslands and marshes are preferred over row crops. In Madagascar, one red owl female avoided closed-canopy forest and concentrated her activities along forest edges (50% of the recorded locations), rice fields (36%) and other cultivated fields (14%). In Australia, at the end of the breeding season, one female masked owl foraged mainly at the interface between forest and sparse woodland, and another radio-tracked individual foraged 23.6% of the time in bushland, 68.9% of the time in residential areas, 7.5% of the time in open country and only rarely in semi-arid environments. One male greater sooty owl foraged predominantly in sandstone gully forest, sandstone ridge woodland including residential land and open water.

A strong preference for some specific hunting habitats during the breeding period may be dictated by the need to capture many prey items during a short period of time. Once the offspring are independent, owls may be less constrained by the strict need to forage highly efficiently and may exploit their territory in a more homogeneous way. Accordingly, during the breeding season, in Germany, six barn owls were found most frequently in open fields (51%), followed by hedges (38%), buildings (4%), gardens and forests (< 5%), whereas in the autumn and winter they spent more time in gardens (29%) and around buildings (28%) and substantially less time in open fields (25%) and hedges (20%). Villages may be more frequently visited in winter than during the breeding period to find a place to hide, get warm and hunt.

## Agricultural, grassland/marsh and forest specialists

Four major categories of Tytonidae can be defined. The majority of barn owls (*Tyto alba* species complex containing the western *T. alba*, the American *T. furcata* and the eastern *T. javanica*) exploit agricultural lands associated with humans. These owls will further expand their distributional range due to deforestation and the spread of introduced small mammals, as observed a hundred years ago in the northern parts of the USA and more recently in Malaysia.

Grass owls, which are found in Africa (*Tyto capensis*), Asia and Oceania (*T. longimembris*), also occur in open land but not necessarily in agricultural fields as they avoid human landscapes. Instead, they are observed hunting in grasslands and marshes. They typically exploit regions where 60- to 200-centimetre-tall dense grasses or sedges homogeneously cover large areas that are usually dry during the breeding season and periodically flooded at other times of the year. Grass owls are therefore habitat specialists and are thus very sensitive to human disturbance as many grasslands and marshes are converted into agricultural fields.

Another category of owls found mainly in Australasia exploits old, undisturbed forests. The lesser sooty owl (*Tyto tenebricosa multipunctata*) and greater sooty owls (*T. t. tenebricosa*, *T. t. arfaki*) live in unlogged rainforests with a tall and dense understorey. Other forest specialists include the Minahassa masked owl (*T. inexspectata*) in Sulawesi, the golden masked owl (*T. aurantia*) on New Britain Island, the red owl (*T. soumagnei*) in Madagascar, and the Seram masked owl (*T. almae*) and the bay owls living in Asia and central Africa (*Phodilus* spp.).

A number of Tytonidae are generalists, exploiting both forests and open landscapes. The Australian and Tasmanian masked owls (*Tyto novaehollandiae novaehollandiae* and *T. n. castanops*) indeed require eucalypt forests for roosting and nesting and forest edge and open woodland for hunting. Other generalists include the Sulawesi masked owl (*T. rosenbergii*) and the Madagascan barn owl (*T. alba hypermetra*).

## FUTURE RESEARCH
- How do conspecifics partition home ranges among each other?
- Are nests usually in the middle of the home range?
- How do males and females partition their territory: do they forage in the same places?
- The core home range should be differentiated from the entire home range, and the territory from the home range.

## FURTHER READING

Almasi, B., Roulin, A. and Jenni, L. 2013. Corticosterone shifts reproductive behaviour towards self-maintenance in the barn owl and is linked to melanin-based coloration in females. *Horm. Behav.* **64**: 161–171.

Arlettaz, R., Krähenbühl, M., Almasi, B., Roulin, A. and Schaub, M. 2010. Wildflower areas within revitalized agricultural matrices boost small mammal populations but not breeding barn owls. *J. Ornithol.* **151**: 553–564.

Brandt, T. and Seebass, C. 1994. *Die Schleiereule: Ökologie eines heimlichen Kulturfolgers*. Wiesbaden: Aula Verlag.

Embar, K., Raveh, A., Burns, D. and Kotler, B. P. 2014. To dare or not to dare? Risk management by owls in a predator–prey foraging game. *Oecologia* **175**: 825–834.

Finck, P. 1990. Seasonal variation of territory size with the little owl (*Athene noctua*). *Oecologia* **83**: 68–75.

Sunde, P. and Bølstad, M. S. 2004. A telemetry study of the social organization of a tawny owl (*Strix aluco*) population. *J. Zool.* **263**: 65–76.

Majestic view of a barn owl.

## 6.4  FLIGHT MECHANICS

# Buoyant and silent

**The barn owl is so silent that even the best microphones can barely detect it in flight. Hearing small mammals but not being heard by them is an absolute necessity for nocturnal birds which hunt on the wing. To float in the air and fly slowly, close to the ground, owls need lift, an upward force that counteracts gravity. They have developed many adaptations to generate this upward buoyant force. The barn owl has been an inspiration for aeronautical engineers.**

Efficient foraging methods are required for barn owl parents to feed so many mouths. Because visual resolution and contrast sensitivity are limited, this nocturnal bird uses acoustic information to locate its prey. As perches are rare in open habitats, sitting and waiting to hear a small mammal wandering about its surroundings is not always an option. One solution is to silently patrol large open fields by flying at low speed (2.5–7 metres per second) close to the ground. Such tactics necessitate the ability to glide and beat the wings slowly (5.6 movements per second) without, however, losing manoeuvrability to quickly turn and strike prey. This is made possible by wings that generate a strong lift, which is twice as high in the barn owl as in the buzzard – although still weaker than in the albatrosses, which effortlessly traverse the Southern Ocean.

The most important adaptation for flying slowly is reduced body mass, which was accomplished by decreasing the size of the small intestine and the sternum area where the two pectoral muscles are attached. Flying slowly just above the ground by gently flapping the wings is relatively effortless with good lift. Hence, the pectoral muscles do not need to be very powerful. In the barn owl, the flight muscles constitute approximately 10% of the body mass, compared with 30% in the pigeons, which must take off very quickly to escape predators.

Enlarged wings are necessary to reduce the load supported by each square centimetre of wing surface. For comparison, barn owl wings are 2.5 times larger than those of similarly sized birds, such as pigeons, and in turn the **wing loading** (ratio of body mass to wing area) is three times lower.

**Generating lift**  To enable the barn owl to become airborne and fly slowly, lift is generated by the air flowing more quickly over than under the wings. The faster air movement on the dorsal side of the feathers reduces the pressure on the wings and hence acts as suction, the same principle which allows planes to fly. The barn owl's highly **cambered wing feathers**, particularly at the tip, are key to this aerodynamic process, especially during low-speed flight – when they act rather like the palms of an Olympic swimmer. To maintain this ability at higher speeds, the **velvet-like surface** of the upper side of the wing ensures stable airflow and increases flying performance. It improves the aerodynamic properties of the feathers by preventing disruption of the airflow.

When flying, pouncing on prey is difficult, requiring rapid changes in velocity (high acceleration/deceleration) and tight turns. This is achieved by the air flowing between the **pliant wing feathers**, which can be actively bent independently of each other. During the upstroke, the resistance to the air that flows between the feathers is attenuated by the flexible properties of the very large inner vane, while the small outer vane reinforces the leading edge.

To transmit the force that acts on the tip of the feathers during flight to the bones, muscles and tendons, a strong anchorage of the feathers to the wing is necessary. This is the function of the very **rigid calamus** (the base of the feather, without barbs), which is of similar length in the barn owl as in the pigeon, although it is proportionally longer in the pigeon, accounting for 23% of the rachis (or shaft, the central part of the feather), compared to 16% in the barn owl. A long, anchored calamus is not necessary in the barn owl because this bird does not flap its wings as frequently or as hard as a pigeon does.

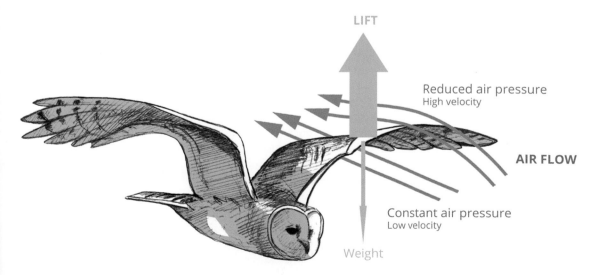

How lift is generated in flight.

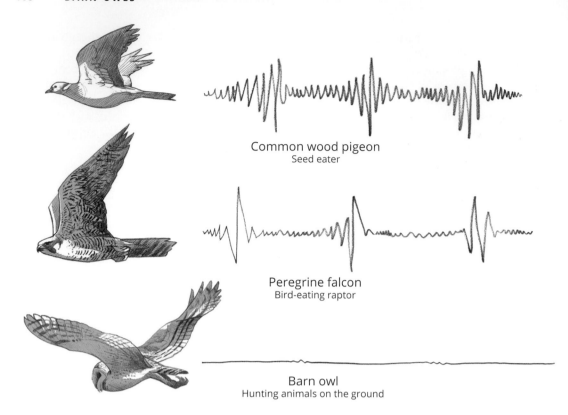

Common wood pigeon
Seed eater

Peregrine falcon
Bird-eating raptor

Barn owl
Hunting animals on the ground

As shown by the sound-wave peaks on the sonograms, the barn owl makes almost no noise in flight, in contrast to the pigeon and peregrine falcon.

**Reducing noise**  Silent flight is just as important for the barn owl as lift and manoeuvrability. The best-known adaptation to reduce noise while flying is the presence of **serrations** on the outer vane of the tenth (outermost) primary wing feather and the tenth greater primary covert. Only these two feathers have serrations, as they are the only ones to form the leading edge of the wing. In the African eagle owl, the tawny owl and the long-eared owl, the serrations extend to the ninth primary, which also forms part of the leading edge. Most nocturnal owls have serrations, except for species such as the tawny fish owls (*Ketupa* spp.), whose submerged prey cannot hear them approaching. Serrations are barbs that separate and bend upwards to form a series of tooth-like shapes. They stabilize the airflow around the wing so that the air is kept close to the wing, not only to dampen noise but also to induce further lift. Because of the increased lift, owls do not need to beat their wings very frequently, which in itself reduces noise by 4–8 decibels, particularly during take-off and landing. Additionally, the velvety surface prevents the barbs of two superposed feathers from coming too close to each other. This reduces noise generated by the air and the feathers themselves when they rub against each other. The need to fly slowly and silently probably explains why the inner vanes of a barn owl's wing feathers are three times wider than those of birds that fly much faster and can afford to be noisy. The high **feather flexibility** (and **porosity**) also helps to dampen noise while flying.

Birds that do not need to reduce noise while flying have sharp edges on the inner and outer vanes of the remiges (flight feathers), whereas nocturnal hunters such as the barn owl have **fluffy fringes at the trailing edge of these feathers**. The barbs of these fringes are not connected, due to a loss of barbules, and thus can roll around the neighbouring feather vanes. In gliding flight the wing needs to form a single structure over which the air flows smoothly, whereas when making tight turns or

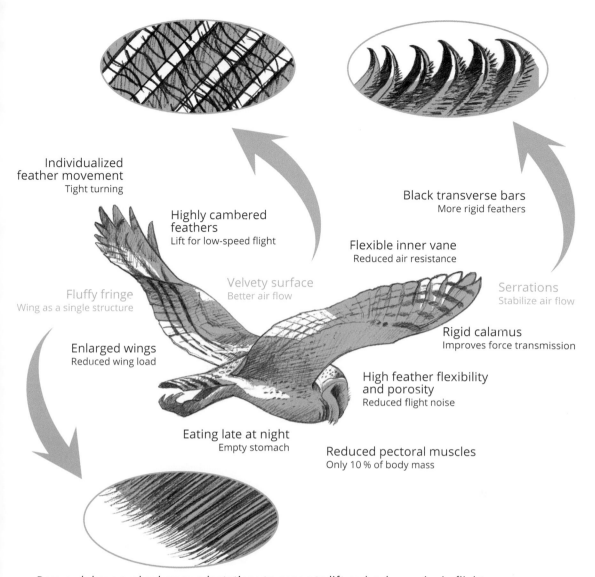

Individualized
feather movement
Tight turning

Highly cambered
feathers
Lift for low-speed flight

Black transverse bars
More rigid feathers

Flexible inner vane
Reduced air resistance

Velvety surface
Better air flow

Fluffy fringe
Wing as a single structure

Serrations
Stabilize air flow

Enlarged wings
Reduced wing load

Rigid calamus
Improves force transmission

High feather flexibility
and porosity
Reduced flight noise

Eating late at night
Empty stomach

Reduced pectoral muscles
Only 10 % of body mass

Barn owls have evolved many adaptations to generate lift and reduce noise in flight.

when taking off, the bird needs to make frequent adjustments, and in that case the wing should not be a single structure. During the upstroke, when the wing feathers separate, the soft fringes also reduce noise by diminishing the turbulence of air flowing through the wing.

**Black transverse bars** In many owls and diurnal raptors, wing and tail feathers have black transverse bars that have attracted little interest. In French barn owls, the proportion of the tenth primary covered by these bars is 8% on average and varies from 0.4% to 20.9%. The function of the melanin-based bars may be to reinforce the feathers, rather like iron bars stabilizing a balcony or beams supporting the roof of a house. Long feathers may need to be reinforced more than shorter ones because they have to withstand greater pressures – and sure enough, longer feathers have more bars. Females are on average heavier than males and accordingly have more bars than males, the bars covering 8.1% and 6.6% of their feathers, respectively. In addition, yearlings have more bars than adults (8.4% vs. 6.4%).

The rigid calamus and the markedly asymmetric inner and outer vanes of a barn owl wing feather.

Details of a barn owl's wing feathers: the serrated leading edge of the tenth primary (*above*) and the velvet-like surface of a feather (*below*). © Michael Todd and Laurent Willenegger

**An inspiration to aeronautical engineers**  Engineers are excited by the ability to fly silently. What we have learned is that wing area, camber, edge shape of the wing and surface texture play critical roles in producing lift and reducing flight noise. Slight changes in these values influence flight characteristics dramatically, and engineers adjust these values to optimize flight performance. This has led to new designs, which do not necessarily require sharp edges and straight wings. For instance, aeroplanes can now be seen with curved wings, with winglets at their tips, and exhaust nozzles with a wavy edge.

## FUTURE RESEARCH

- The amount of black transverse bars should be measured in a range of *Tyto* taxa, to identify the key biogeographical variables that could have promoted their evolution.
- The next steps in flight biology should be to investigate dynamic processes such as flapping flight and active wing-shape control. This includes a study of how owls react under turbulence to actively reduce flight noise.
- Modern 3D reconstruction software and high-speed imaging might help to reconstruct the 3D shape of a barn owl's wing during flight. Simulation software could then be used to study the air-flow phenomena around the morphing wing.
- Using new technologies, such as accelerometers placed along with GPS devices, will provide the opportunity to determine how barn owls alternate wing flapping and gliding in relation to various factors including body mass and landscape features.

## FURTHER READING

Bachmann, T. and Wagner, H. 2011. The three-dimensional shape of serrations at barn owl wings: towards a typical natural serration as a role model for biomimetic applications. *J. Anat.* **219**: 192–202.

Bachmann, T., Klän, S., Baumgartner, W., Klaas, M., Schröder, W. and Wagner, H. 2007. Morphometric characterisation of wing feathers of the barn owl *Tyto alba pratincola* and the pigeon *Columba livia. Front. Zool.* **4**: 23.

Bachmann, T., Mühlenbruch, G. and Wagner, H. 2011. The barn owl wing: an inspiration for silent flight in the aviation industry? *Proc. SPIE – Bioinspiration, Biomimetics, and Bioreplication* 7975.

Bachmann, T., Emmerlich, J., Baumgartner, W., Schneider, J. M. and Wagner, H. 2012. Flexural stiffness of feather shafts: geometry rules over material properties. *J. Exp. Biol.* **215**: 405–415.

Bachmann, T., Wagner, H. and Tropea, C. 2012. Inner vane fringes of barn owl feathers reconsidered: morphometric data and functional aspects. *J. Anat.* **221**: 1–8.

Barton, N. W. H. and Houston, D. C. 1994. Morphological adaptation of the digestive tract in relation to feeding ecology of raptors. *J. Zool.* **232**: 133–150.

Roulin, A., Mangel, J., Wakamatsu, K. and Bachmann, T. 2013. Sexually dimorphic melanin-based color polymorphism, feather melanin content and wing feather structure in the barn owl (*Tyto alba*). *Biol. J. Linn. Soc.* **109**: 562–573.

A barn owl pouncing on its prey.

## 6.5   HUNTING METHODS

# Tora! Tora! Tora!

**Barn owls employ a suite of different hunting methods to varying extents – flying slowly close to the ground, sitting on a perch and waiting for passing prey, or hovering like a kestrel. Because of the high risk of predation, small mammals have evolved adaptations to escape their predators. In turn, the arms race between barn owls and their prey has promoted the evolution of refined hunting tactics adjusted to each specific condition, including surprise attacks that give prey little chance to escape.**

Barn owls can use different hunting techniques. The large and heavy sooty and masked owls forage mainly by sitting on a perch and waiting for prey to pass. This sit-and-wait method is a typical 'low-cost/low-benefit' strategy. Waiting on a post is energetically cheap but the time needed to finally capture a small mammal can be long. The kestrel-like hovering method, in contrast, is a 'high-cost/high-benefit' strategy. Following a wandering small mammal by beating the wings to remain in one position like a helicopter is particularly tiring, but the likelihood of capturing the prey is high. Foraging by roaming just above the ground at a low speed is a good compromise: it is more efficient than sit-and-wait and less costly than hovering flight.

Grass owls hunt mainly in flight, whereas barn owls can use all three methods (although the kestrel-like hovering is rarely observed). An even rarer method is to flap the wings against bushes to scare roosting passerines. Whichever technique is employed, Tytonidae prefer to attack their prey by surprise, either on the wing or by quietly sitting on a perch, rather than by actively chasing their prey.

The barn owl's ability to adapt to different contexts and environments is striking. The anatomy of the barn owl is particularly well adapted to patrolling on the wing and hunting from a perch, but less suitable for hovering like a helicopter. While sitting on a pole, the owl can rotate its head and track prey with its acute hearing and vision, and its flight is silent. These two foraging strategies allow the barn owl to launch 'surprise attacks', whereas when hovering it is likely to generate enough turbulence to be detected by the prey.

The barn owl's propensity to forage on the wing is highly variable and possibly population-specific. For example, researchers have reported owls hunting by flying in California and by perching in Texas. This most likely depends on environmental parameters and habitat characteristics such as weather and the availability of perches.

**To fly or to sit?** Barn owl parents must bring a lot of food to the family over a short period of time, and to this end they frequently hunt on the wing. In winter, when there is no need to capture copious amounts of prey each night, barn owls can opt to use the sit-and-wait tactic more frequently. However, when prey is scarce during the winter, given low rodent reproduction, barn owls may be more active, hunting in flight. During the breeding season Scottish owls hunting by day did so from a perch in 54% of cases, compared to 87% of cases on dark winter nights. The frequent use of perch hunting in Scotland may be due to the short nights during the summer breeding season, which cause barn owls to extend foraging into the very long days. This is not the case in Switzerland, where breeding males searched for prey on the wing on 22 occasions (92%) and in only two cases (8%) did so by sitting on a perch during the 7- to 8-hour-long nights. This raises the interesting possibility that Scottish owls sleep less, using the cheap sit-and-wait hunting method, whereas on the European mainland barn owls must rest for longer because they use a more expensive foraging technique.

Small rodents are scared by a flying owl but are less likely to notice the presence of an owl sitting quietly on a perch. Hence, there might be environments in which barn owls will fare better using the sit-and-wait method. Rain is noisy, reducing the owl's ability to detect its prey, and porous wet feathers become too heavy to fly, which may cause owls to sit. Additionally, vegetation can restrict free flight, and if perches are abundant it may be more cost-efficient to use them for hunting. This may be the case for short-winged barn owls, such as those on the Galápagos Islands, where, over 8.5 hours of observations, owls hunted from a perch 72% of the time.

However, not all perches are suitable. Owls must be close enough to the ground to clearly see small mammals but far enough away to avoid detection. This explains why barn owls have been observed using perches of varying heights in different regions – in India they preferentially use perches 2.7 metres high (instead of 2.1 and 1.5 metres), whereas in Scotland they perch at 1–1.5 metres, and in Malaysia they use taller perches of 3–5 metres to capture large rats.

**Launching an attack** Before launching an attack, a barn owl must assess the distance between its take-off point and the exact location of a small mammal.

**The closer the better.** In aviaries, successful attacks are launched at a mean distance of 0.57 metres, and unsuccessful attacks at 0.71 metres. These distances are so short that flight speed (4.5 metres per second on average) is not a key factor in determining the success of an attack.

**The less active the prey is, the better.** A total of 90% of attacks on stationary prey are successful, compared with only 21% of attacks on moving prey. This is why most voles are attacked when they cease to move or just after they start to move again, but not in the middle of a movement.

**Anticipate the clash point.** Owls do not hastily launch an attack. They fix their eyes on the prey to estimate the direction in which the small mammal is likely to move, to anticipate the likely position of the clash point. If during the attack the stationary mouse suddenly moves, the owl has a 50–55% chance of capturing it if the movement is directly away from the owl, 18–37% if the mouse moves towards the owl, and 23% if sideways.

**In-flight correction.** Prey flees at the last moment in half of barn owl attacks. If the escape is perpendicular to the axis of the attack, the owl must perform a very difficult manoeuvre to adjust its path to that taken by the prey, and it is generally forced to land and launch a second attack. If the owl misses a rodent that is moving directly away from or towards it, it adjusts its flight path to catch the moving prey in 70% of cases and otherwise returns to its roost (23% of cases) or lands on the spot where the prey was initially located (7% of cases). The decision to pursue an attack or to give up depends on when

A sequence showing a barn owl using the sit-and-wait foraging method. After Warren Photographic.

the prey tries to escape. Barn owls can successfully adjust their flight path if the escape movement of the prey occurs soon after the attack is launched, but not if it happens when the owl is very close to the prey.

**Adjust hunting technique to prey species.** Israeli barn owls use different methods for capturing voles and spiny mice, which exhibit different behaviour in front of predators. Mice flee in response to an owl to look for refuge, whereas voles alternate freezing (to avoid making noise, relying on the camouflage of their brown fur) and fleeing (in case the owl has spotted their location). Thus, when an owl misses a mouse, it continues to chase it, but when an owl misses a vole, it goes back to its roost and waits until the vole resumes activity. This behaviour explains why in captivity it takes 16 minutes to capture a mouse but 31 minutes to catch a vole.

## Capturing, killing, eating

How barn owls kill their prey has only been studied in captive Italian barn owls. On 54 occasions, they landed directly on the prey, most often using one foot (67% of the time), and on seven occasions landed close to the prey and pursued it on foot before grasping it. They then seized the prey on the head usually with two feet and killed it by twisting its neck, rather than with a beak strike as in the little owl, tawny owl and long-eared owl.

## FUTURE RESEARCH

- The environmental factors and physiological conditions under which barn owls use the sit-and-wait or flying hunting methods should be investigated. Descriptive studies are also required to examine whether these methods are used at different times of the day and night, and whether their frequency varies geographically.
- Given the pronounced variation in plumage colouration between individuals, do dark reddish and white individuals differ in their success with in-flight hunting, and in their propensity to use this technique?
- The success rate of an attack is almost 100% when prey is stationary, but strongly decreases when prey is fleeing. Thus, capture success mainly depends on whether the prey animal notices the owl, which suggests that the camouflage of the owl could be very important. Studying the role of variation in plumage colouration on the probability that prey notice a barn owl would provide important information about predator–prey interactions.

## FURTHER READING

Abramsky, Z., Strauss, E., Subach, A., Kotler, B. P. and Riechman, A. 1996. The effect of barn owls (*Tyto alba*) on the activity and microhabitat selection of *Gerbillus allenbyi* and *G. pyramidum*. *Oecologia* **105**: 313–319.

Brown, J. S., Kotler, B.P., Rosemary, J. S. and Wirtz, W. O. 1988. The effects of owl predation on the foraging behavior of heteromyid rodents. *Oecologia* **76**: 408–415.

Csermely, D. and Sponza, S. 1995. Role of experience and maturation in barn owl predatory behavior. *Boll. Zool.* **62**: 153–157.

Csermely, D., Casagrande, S. and Sponza, S. 2002. Adaptive details in the comparison of predatory behaviour of four owl species. *Ital. J. Zool.* **69**: 239–243.

Edut, S. and Eilam, D. 2004. Protean behavior under barn-owl attack: voles alternate between freezing and fleeing and spiny mice flee in alternating patterns. *Behav. Brain Res.* **15**: 207–216.

Fux, M. and Eilam, D. 2009. How barn owls (*Tyto alba*) visually follow moving voles (*Microtus socialis*) before attacking them. *Physiol. Behav.* **98**: 359–366.

Fux, M. and Eilam, D. 2009. The trigger for barn owl (*Tyto alba*) attack is the onset of stopping or progressing of the prey. *Behav. Processes* **81**: 140–143.

Hausmann, L., Platchta, D. T. T., Singheiser, M., Brill, S. and Wagner, H. 2008. In-flight corrections in free-flying barn owls (*Tyto alba*) during sound localization tasks. *J. Exp. Biol.* **211**: 2976–2988.

Ilany, A. and Eilam, D. 2008. Wait before running for your life: defensive tactics of spiny mice (*Acomys cahirinus*) in evading barn owl (*Tyto alba*) attack. *Behav. Ecol. Sociobiol.* **62**: 923–933.

Shifferman, E. and Eilam, D. 2004. Movement and direction of movement of a simulated prey affect the success rate in barn owl *Tyto alba* attack. *J. Avian Biol.* **35**: 111–116.

A barn owl eating a mouse.

## 6.6 PREY SELECTION

# The chase may be better than the catch

**Do barn owls select their prey, or do they capture them opportunistically in proportion to their availability? Their strategy is to exploit selected habitats where the probability of catching specific prey is high. Although any small mammal that the owl encounters may be captured, some species are more often preyed on than others, owing to their highly vulnerable behaviour. However, even if the probability of successfully catching the available prey is similar, barn owls selectively choose some prey over others. Hence, when food is plentiful, can barn owls be choosy and pick the meal they like best?**

Predators should limit the costs of hunting by reducing the amount of time taken to encounter a prey animal, the number of capture attempts and the handling time to consume it. When flying relatively quickly above the ground, barn owls have little time to decide whether an encountered prey is worth the catch. The best strategy is therefore to forage in their preferred open habitats where their staple prey is abundant and to capture any potential prey animal they meet. Once an animal is captured, the owl can decide what to do next. For instance, it might not be worth transporting a small shrew to the nest over a long distance, as observed in one German study. The owl should rather eat it on the spot before resuming hunting. On the other hand, if a prey animal is too large to be easily eaten, nestlings may neglect it and wait for parents to bring smaller and easier-to-handle ones. Large prey items are sometimes wasted, because they quickly rot under high ambient temperatures.

**An opportunistic gourmand**   When the staple prey species becomes more abundant, it usually represents a larger portion of the barn owl's diet. Over a period of eight years in France, the proportion of the diet composed of the common vole was directly related to its abundance. An increase in its availability also led to an increase in the proportion of water voles in the diet, which indicates that when barn owls search for common voles, they also encounter water voles. In contrast, the proportion of red-toothed shrews in the diet increased only when voles became less abundant. Thus, only when the commonest prey becomes scarce do owls capture other species that are otherwise neglected even if relatively abundant. The inclusion of less commonly eaten species in the diet of French barn owls does not depend on their own abundance but rather on the abundance of the most profitable prey species, the common vole.

Not all prey is equally vulnerable. In one study in Argentina, owls preyed on large rodents, and in one study in Spain on large male marsh frogs, at higher rates than the abundance of these prey in the field would predict. When taking care of their offspring in a hidden burrow, female rodents are out of reach from predators, which explains why in Argentina and Wales they are captured less often during the breeding season than in autumn and winter. Interestingly, whereas mature rodents are disproportionately captured in Chile, juveniles are more often preyed on in Brazil. Finally, western Mediterranean mice with bone deformations are particularly susceptible to predation in Spain.

What causes these non-random patterns of predation? The more frequent capture of some prey than predicted by their abundance occurs often because their behaviour makes them particularly vulnerable. Under controlled aviary conditions, captive American barn owls caught twice as many meadow voles as white-footed and deer mice, which are more active and hence more difficult to catch. In Australia, juvenile female house mice were particularly exposed to barn owl predation because adults force them out of the more protected areas in habitats where owls preferentially hunt. Small individuals could also suffer from their inability to escape very rapidly, males from their conspicuous sexual displays and juveniles from their inexperience and their tendency to wander outside of dense vegetation when dispersing.

The above explanations lead us to believe that barn owls do not capture prey selectively but rather opportunistically. Is this really the case? Even if barn owls are opportunistic predators, usually capturing prey in proportion to the prey's availability and vulnerability, there are situations where owls may neglect some potential prey and wait for another. Detailed observations in aviaries show that captive barn owls select their prey when offered the choice. In India, they preferentially captured prey of medium weight rather than lighter and heavier prey. In Germany, they consumed dead and live mice weighing 10 grams rather than 3 grams; when offered dead animals, they preferred individuals weighing 40 grams rather than 10 grams; and when offered live rats, they captured and devoured those weighing 40 grams but neglected those weighing 160 grams. Large rats may be appealing, but they are difficult not only to subdue, as they defend themselves, but also to handle for eating.

To conclude, although barn owls are usually opportunist hunters, they can, if given the choice, choose which prey to capture.

## FUTURE RESEARCH
• Do nestlings have food preferences? This could be tested by simultaneously offering a set of different prey species.

## FURTHER READING

Bellocq, M. I. 1998. Prey selection by breeding and nonbreeding barn owls in Argentina. *Auk* **115**: 224–229.

Bernard, N., Michelat, D., Raoul, F., Quéré, J.-P., Delattre, P. and Giraudoux, P. 2010. Dietary response of barn owls (*Tyto alba*) to large variations in populations of common voles (*Microtus arvalis*) and European water voles (*Arvicola terrestris*). *Can. J. Zool.* **88**: 416–426.

Derting, T. L. and Cranford, J. A. 1989. Physical and behavioral correlates of prey vulnerability to barn owl (*Tyto alba*) predation. *Am. Midl. Nat.* **121**: 11–20.

Dickman, C. R., Predavec, M. and Lyman, A. J. 1991. Differential predation of size and sex classes of mice by the barn owl, *Tyto alba*. *Oikos* **62**: 67–76.

Embar, K., Mukherjee, S. and Kotler, B. P. 2014. What do predators really want? The role of gerbil energetic state in determining prey choice by barn owls. *Ecology* **95**: 280–285.

Ille, R. 1991. Preference of prey size and profitability in barn owls *Tyto alba guttata*. *Behaviour* **116**: 180–189.

Pribbernow, M. 1996. Nahrungsökologische Untersuchungen an Schleiereulen (*Tyto alba*, Scopoli 1769) in der Uckermark. Diplom-Arbeit, Humboldt Universität, Berlin.

Roulin, A. 2004. The function of food stores in bird nests: observations and experiments in the barn owl *Tyto alba*. *Ardea* **92**: 69–78.

Vanitha, V. and Kanakasabai, R. 2009. Prey selection by the barn owl *Tyto alba* (Scopoli, 1769) in captivity. *J. Threat. Taxa* **1**: 361–365.

The barn owl diet is mainly composed of small mammals.

## 6.7  DIET

# Foodie

**Small mammals constitute the staple prey of barn owls. These are primarily small rodents, with species in the Muridae family being found in the diet of the barn owl across all continents. Insectivorous mammals are captured in large numbers in Asia and Europe, and marsupials in low numbers in the Americas and Oceania. The diet of other Tytonidae (sooty owls, masked owls and African grass owls) is more diverse than the diet of barn owls and eastern grass owls. In Europe, the abundance of invertebrates and insectivorous animals (shrews, moles, birds, bats) has decreased over the last century.**

Containing the remains of up to fifteen small mammals, a pellet can be impressively long, up to 11 centimetres. Pellets are amazing sources of information about the diet of a predator and its ecology, and can also provide valuable information about mammalian communities in the region.

### Diet of the barn owl species complex
Worldwide, the barn owl diet consists of 72–99% mammals, with the rest being birds, invertebrates, amphibians and reptiles. Based on an extensive analysis of pellets containing a total of 4.8 million identified small mammals, rodents (mainly the mouse-like Muridae and the vole-like Cricetidae, Heteromyidae and Nesomyidae) are the most commonly captured prey, followed by insectivorous small mammals (mainly shrews, Soricidae) and marsupials in Australia and South America. While Muridae are found in large numbers in the barn

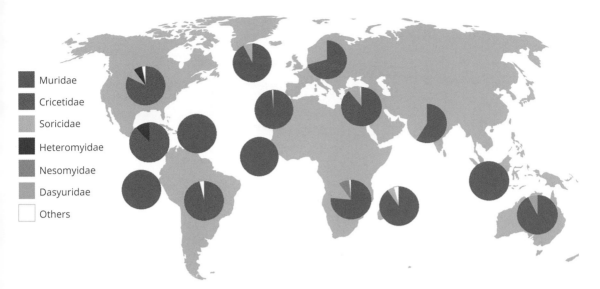

Muridae
Cricetidae
Soricidae
Heteromyidae
Nesomyidae
Dasyuridae
Others

Frequency of the most abundant small mammalian families as prey in the barn owl species complex. This figure is based on 1240 studies published between 1860 and 2016 and totalling 4 769 542 identified small mammals.

owl's diet worldwide, Cricetidae are captured only in the Americas, Europe and the Middle East. Soricidae are caught in Europe, North Africa, the Middle East and North America.

The only family of prey captured on all continents is Muridae, with mice of the genus *Mus* being particularly abundant in Australia, North Africa, the Middle East, the Iberian Peninsula and Madagascar. Rats of the genus *Rattus* are also found on all continents, with high numbers being consumed in Southeast Asia, the Caribbean islands, the northern parts of South America and Madagascar.

**Shannon's diversity index** indicates the extent to which an animal's diet is diversified, accounting for both number of species and frequency counts. The more specialist the diet the closer to 0 is the index value. Using this index, the barn owl's diet appears to be more diverse on continents than on islands (on average 1.44 and 1.17, respectively), and more diverse in South America and continental Europe than in the Middle East, Africa, North America, Asia, and Australia – as shown in the table on page 120 and the maps on page 121. The barn owl diet is particularly diverse in South Africa, the centre of North America, the southern parts of South America and eastern Europe. In Australia, North and West Africa, and the northern parts of North and South America, the diet is more specialized, concentrated on a few key species.

**Europe**  A total of 907 papers reported 4.54 million prey items, and detailed analysis of these studies showed pronounced temporal changes in diet composition from 1860 to 2012. Rodents are by far the most frequent prey in Europe, accounting, on average, for 71.6% of mammalian prey. Insectivorous small mammals comprise the second largest group (28.4% without counting bats), including seven species of red-toothed shrews (the median percentage of small mammals that are *Sorex* is 12.5%), six species of white-toothed shrews (*Crocidura*, 7.7%), the Etruscan shrew (*Suncus etruscus*, 2.5%), two species of water shrews (*Neomys*, 0.4%) and six species of moles (*Talpa*, less than 0.0001%). The consumption of these insectivores decreased during the last century, as did the consumption of bats, which are more often depredated in eastern than in western Europe, in southern than in northern Europe, and on islands. Birds are captured relatively often, particularly on islands and in southern and eastern Europe. The consumption of birds, particularly house sparrows (65.7% of all captured birds), also declined over the last century, particularly in northern and eastern Europe. The probable cause of the temporal decline in the abundance of shrews, moles, bats and

Percentage of various prey types and Shannon's diversity index in the barn owl species complex and in the grass owls, Australian masked owl and sooty owl. Only studies in which researchers did not restrict diet analyses to mammals were considered, which explains the differences in sample sizes between this table and the text. The data were obtained from scientific papers published between 1860 and 2016.

| | Barn owl species complex (*Tyto alba, T. furcata, T. javanica*) | | | | | | | | |
| | Europe | UK | Africa | Madagascar | Middle East | North America | South America | Asia | Australia |
|---|---|---|---|---|---|---|---|---|---|
| Number of prey | 3193542 | 609191 | 87578 | 6360 | 34447 | 332546 | 76448 | 17480 | 15370 |
| Number of studies | 493 | 52 | 54 | 5 | 25 | 111 | 64 | 9 | 17 |
| Shannon's diversity index | 1.9 | 1.76 | 1.69 | 1.13 | 1.72 | 1.57 | 1.9 | 1.35 | 0.98 |
| ***Mammals*** | | | | | | | | | |
| Rodents | 68.8 | 72.31 | 68.9 | 58.78 | 66.23 | 89.6 | 82.05 | 79.2 | 67.48 |
| Insectivores | 27.46 | 26.45 | 7.64 | 5.13 | 21.69 | 7.82 | 0.05 | 16 | |
| Marsupials | | | | | | 0.01 | 0.9 | | 6.06 |
| Primates | | | 0.23 | 1.13 | | | | | |
| Carnivores | 0.01 | | | | | 0.01 | | | |
| Bats | 0.14 | 0.03 | 0.32 | 7.1 | 1.56 | 0.14 | 1.62 | 1 | 0.12 |
| ***Non-mammals*** | | | | | | | | | |
| Birds | 2.64 | 1.04 | 13.29 | 6.25 | 6.08 | 2.15 | 4.5 | 1.52 | 3.33 |
| Reptiles | 0.05 | 0.003 | | | 0.48 | 0.02 | 0.64 | 0.15 | 16.85 |
| Amphibians | 0.55 | 0.13 | 2.31 | 19 | 0.08 | 0.005 | 2.69 | 0.22 | 1.98 |
| Invertebrates | 0.35 | 0.02 | 7.31 | 3.57 | 3.88 | 0.25 | 7.54 | 1.95 | 4.17 |
| Crustaceans | 0.00003 | | | | | 0.01 | | | |
| Fish | 0.00059 | 0.0003 | | | | 0.01 | 0.02 | | |
| Gastropods | 0.00059 | | | | | | | | |

| | Other Tytonidae | | | |
| | Grass owl | | Masked owl | Sooty owl |
| | *Tyto capensis* | *Tyto longimembris* | *Tyto novaehollandiae* | *Tyto tenebricosa* |
|---|---|---|---|---|
| Number of prey | 1666 | 1117 | 2562 | 7797 |
| Number of studies | 8 | 9 | 9 | 11 |
| Shannon's diversity index | 2.26 | 1.39 | 2.01 | 2.21 |
| ***Mammals*** | | | | |
| Rodents | 91.24 | 88.63 | 61.59 | 18.47 |
| Insectivores | 7.02 | 2.78 | | |
| Marsupials | | 1.97 | 27.71 | 79.29 |
| Primates | | | | |
| Carnivores | | | 0.04 | |
| Bats | | | 0.31 | 0.1 |
| ***Non-mammals*** | | | | |
| Birds | 1.74 | 4.21 | 7.65 | 1.72 |
| Reptiles | | 0.09 | 0.16 | 0.33 |
| Amphibians | | | | |
| Invertebrates | | 2.33 | 2.54 | 0.08 |
| Crustaceans | | | | 0.01 |
| Fish | | | | |
| Gastropods | | | | |

*Opposite page, top and middle maps:* Proportion of the barn owl diet composed of mice of the genus *Mus* and rats of the genus *Rattus*. For example, in Australia, the barn owl diet can comprise up to 86% mice. The data were obtained from scientific papers published between 1860 and 2016.

*Opposite page, bottom map:* Shannon's diversity index indicates how diverse the diet is in the barn owl species complex. High values (in red) indicate that the barn owl's diet is composed of many prey species captured at approximately equal frequencies, whereas low values (in blue) indicate that the diet is composed of few prey species. This figure is based on 1274 papers published between 1860 and 2016 and considers only small mammalian prey.

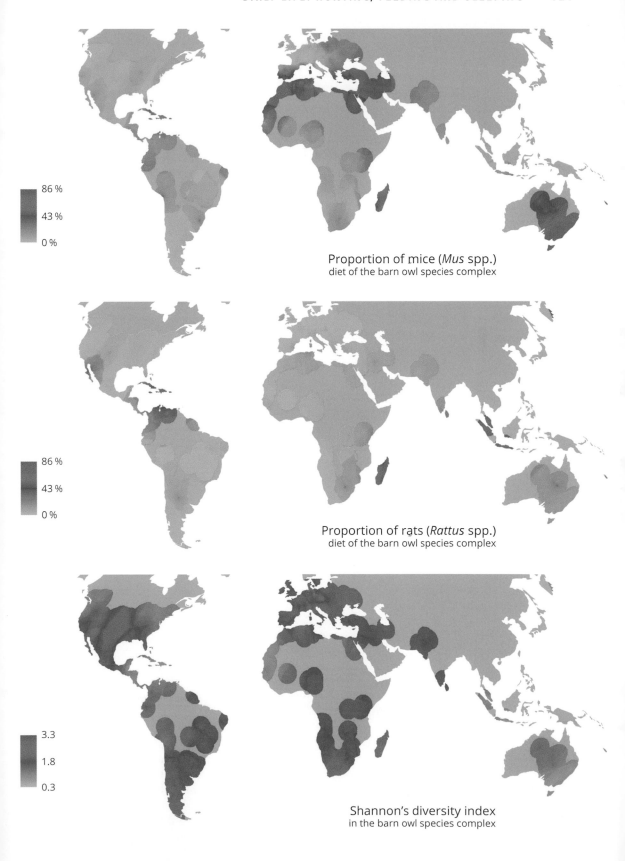

Proportion of mice (*Mus* spp.)
diet of the barn owl species complex

Proportion of rats (*Rattus* spp.)
diet of the barn owl species complex

Shannon's diversity index
in the barn owl species complex

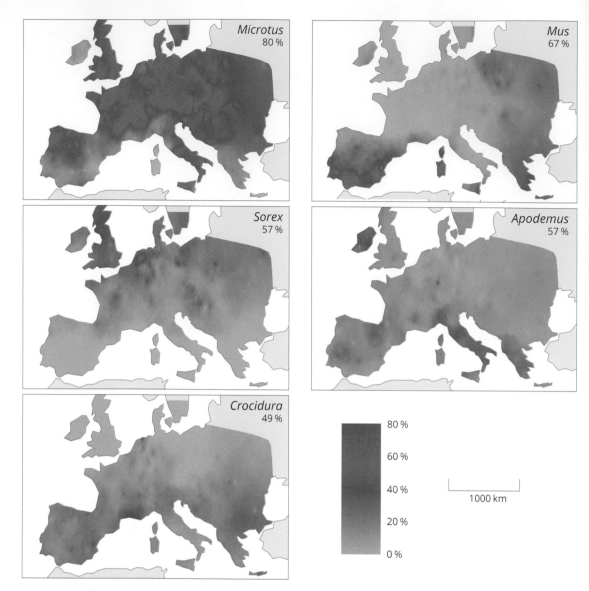

Frequency distribution of voles (fifteen species of *Microtus*), mice (four species of *Mus*), wood mice (six species of *Apodemus*), red-toothed shrews (seven species of *Sorex*) and white-toothed shrews (six species of *Crocidura*) in the diet of European barn owls. This is based on 897 scientific papers published between 1860 and 2016.

birds in the barn owl diet is the strong decline in the populations of the invertebrate species forming their prey base.

European barn owls capture few diurnal reptiles. Amphibians can be quite active on rainy nights and are thus captured ten times more often than reptiles by the nocturnal predators. Interestingly, barn owls avoid the toxic Bufonidae, Alytidae and Salamandridae. Gastropods, crayfish and fish are rarely consumed.

**North America**  A total of 124 papers reported 350 160 prey items. The most commonly captured family of small mammals in North America is Cricetidae, with voles (*Microtus* spp.) representing 38.1% of all small mammals and being particularly abundant in the northern parts of the

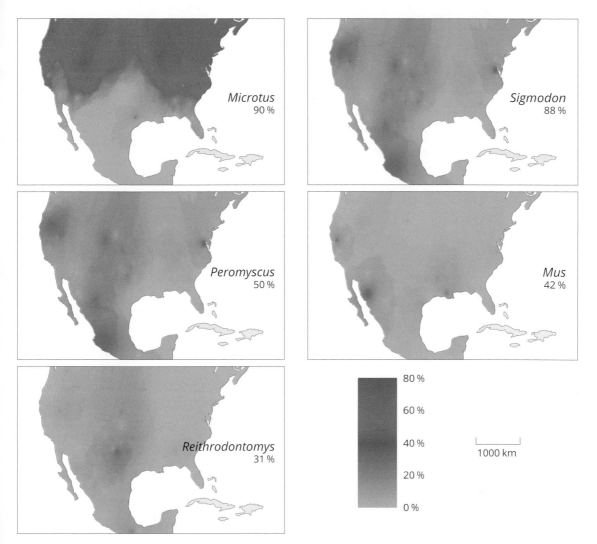

Frequency distribution of voles (nine species of *Microtus*), New World rats and mice (six species of *Sigmodon*), deer mice (eight species of *Peromyscus*), mice (two species of *Mus*) and New World harvest mice (six species of *Reithrodontomys*) in the diet of North American barn owls. This is based on 124 scientific papers published between 1924 and 2011.

continent. The second most abundant genus is the New World rats and mice (*Sigmodon*, 10.6% of small mammals), particularly along the Gulf of Mexico. The groove-toothed New World harvest mice (*Reithrodontomys*, 3.9%) are found mainly in the centre of the continent. Other families comprising more than 1% of the mammalian prey are deer mice (*Peromyscus*, 9.8%), semi-aquatic rodents (*Oryzomys*, 6.2%), pocket mice (*Perognathus*, 3.3%; *Chaetodipus*, 2.8%), kangaroo rats (*Dipodomys*, 3.1%), mice (*Mus*, 3.6%), large shrews (*Blarina*, 2.9%), small shrews (*Cryptotis*, 2.6%) and red-toothed shrews (*Sorex*, 1.5%).

**South America**  A total of 75 studies reported 95 551 prey items. Compared to other continents, invertebrates are more often captured in South America. Among small mammalian prey, the family Cricetidae forms the bulk of the diet (87.7%) followed by Muridae (8.3%) and Didelphidae (1.4%), while each of the other families (Ctenomyidae, Caviidae, Abrocomidae, Octodontidae, Echimyidae,

A scary view for a rodent.

Microbiotherlidae) represents less than 1% of the diet. The most frequently captured species are the flavescent colilargo (*Oligoryzomys flavescens*, 12.2% among all mammalian prey), the Azara's akodont (*Akodon azarae*, 8.7%), the house mouse (*Mus musculus*, 6.2%), the small vesper mouse (*Calomys laucha*, 5.6%), the long-tailed colilargo (*Oligoryzomys longicaudatus*, 3.1%), the Darwin's leaf-eared mouse (*Phyllotis darwini*, 2.6%) and the bunny rat (*Reithrodon auritus*, 2.6%).

**Africa**   A total of 76 papers reported 218 209 prey items. The most commonly captured family is Muridae, particularly in North Africa. Two other families often consumed are the Soricidae and Nesomyidae. Other families include Bathyergidae, Chrysochloridae, Dipodidae, Erinaceidae, Gliridae, Macroscelididae and Spalacidae. Compared to other continents, birds and invertebrates are captured in large numbers.

**Middle East**   A total of 32 studies reported 41 220 prey items. Middle Eastern barn owls have a diet similar to that of European owls, except that the frequency of Muridae, particularly mice (*Mus* spp.), is much higher.

**Asia**   A total of 11 studies reported 19 408 prey items in Asia. The proportion of Muridae is even higher in Asia than in the Middle East, particularly on islands located in southeastern Asia, where three studies in Java and Indonesia found mainly rats in the barn owl diet.

**Australia**   A total of 20 studies reported 17 620 prey items. Severe mammal extinctions and declines over the last 200 years since European settlement on the continent have led to the disappearance of historically important species from the barn owl's diet. Data from studies of sub-fossil pellet deposits in cave roosting sites demonstrate this. Introduced species, primarily the house mouse (*Mus musculus*), now dominate the diet in many areas. Studies in recent decades reveal, unsurprisingly,

A barn owl swallowing its rodent prey.

that Muridae make up the bulk of the diet with five genera (*Leggadina forresti, Mus musculus*, four species of *Notomys*, four species of *Pseudomys*, four species of *Rattus*). This is followed by five genera of Dasyuridae (marsupials, with two species of *Antechinomys*, two species of *Antechinus*, four species of *Sminthopsis*, five species of *Planigale*, and *Ningaui ridei*), the eastern pygmy possum (*Cercartetus nanus*) and many reptiles.

**Diet of other Tytonidae**  The study of the diets of barn owls and their relatives can provide important information about past biodiversity changes. Europeans settled in Australia in 1788 and in Tasmania in 1803, leading to dramatic changes in the local flora and fauna. Currently, masked owls consume mainly introduced animals, which constitute up to three-quarters of their diet. Diet impoverishment arising from the arrival of European settlers was therefore probably even more drastic in masked owls and sooty owls than in eastern barn owls.

Comparison of the prehistoric and present-day diets of the sooty owl, thanks to the analysis of sub-fossil pellet deposits accumulated in caves, has revealed that more than 75% of the original prey species have declined. This large owl lives in forests and hence has a diet markedly different to that of other barn owls, consisting of marsupials, including large possums and gliders, weighing up to 1.5 kilograms. The Australian and Tasmanian masked owl is also very large but hunts in open landscape and woodland like the barn owls. Like the sooty owl, the masked owl's diet contains many more marsupials than the diet of the barn owl. Grass owls in Africa and Australasia consume many rodents, but in Australasia their diet is not diverse. In contrast, the sooty owls, masked owls and African grass owls have diets that are more diverse than barn owls (see table on page 120).

## FUTURE RESEARCH
- The complete list of all prey species (including birds and invertebrates) should be published even if researchers are mainly interested in a particular group of prey, such as small mammals.
- Many researchers analyse barn owl pellets, but most data are not accessible. Because of a lack of data over large geographic areas, maps of prey abundance could be drawn only for Europe and North America.

## FURTHER READING
Bilney, R. J., Cook, R. R. and White, J. 2006. Change in the diet of sooty owls (*Tyto tenebricosa*) since European settlement: from terrestrial to arboreal prey and increased overlap with powerful owls. *Wild. Res.* **33**: 17–24.

Roulin, A. 2016. Strong decline in the consumption of invertebrates by barn owls from 1860 to 2012 in Europe. *Bird Study* **63**: 146–147.

Roulin, A. 2016. Shrews and moles are less often captured by European barn owls nowadays than 150 years ago. *Bird Study* **63**: 559–563.

Roulin, A. and Christe, P. 2013. Geographic and temporal variation in the consumption of bats by European barn owls. *Bird Study* **60**: 561–569.

Roulin, A. and Dubey, S. 2012. The occurrence of reptiles in barn owl diet in Europe. *Bird Study* **59**: 504–508.

Roulin, A. and Dubey, S. 2013. Amphibians in the diet of European barn owls. *Bird Study* **60**: 264–269.

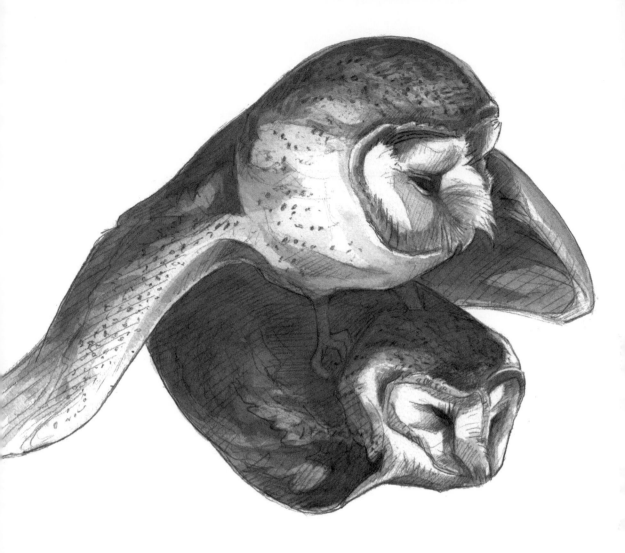

# 7 Sexual behaviour

7.1    Courtship and copulation
7.2    Mating system
7.3    Fidelity and divorce

Courtship feeding in the barn owl, with a male offering a prey item to his mate.

# Flirt

**Barn owls have a demanding sexual life. In Europe, they search for a nesting site and a mate in early winter, and in northern temperate regions they can copulate from February to November. The most sexually active individuals may therefore dedicate most of the year to activities related to reproduction. Sex for barn owls is ceremonial: males offer food to court females, they vocalize during copulation, and when eggs are being laid they can copulate more than once an hour. Barn owls may therefore copulate hundreds of times just to produce a single clutch.**

The sexual behaviour of barn owls in Europe has been described in great detail.

**Mating season**  European barn owls, particularly the more experienced older adults, prospect nest sites in winter. Males defend a territory, and females sample sites and mates before deciding where they will settle. A female barn owl may leave many rejected suitors in her wake. Males make advances to approach females by offering prey items, and females can refuse them and leave the site without consuming these items, which decompose and finally dry out. In Switzerland, 240 nest boxes contained a clutch, and 27 others contained only mummified prey, implying that at least 10% of courtships did not result in a clutch. In one site, we even found 76 rotten prey items. Because barn owls do not store food, as jays or squirrels do in forests, these prey items were offered in vain. This often occurs in good breeding years, when all sexually mature individuals are trying to reproduce. In such years, a

A male barn owl offering a prey item before copulating. Photos were taken with a camera trap. ©
Robin Séchaud & Kim Schalcher

male may attract several females before finding his soul mate. There are so many males attempting to
breed that females can take quite a bit of time to sample males before choosing a mate.

Males attract a mate by emitting up to 500 calls per night and flying in circles around their nest
cavity. They invite females to come inside their nests, where they scratch the ground, as if to show
her where eggs could be laid. Paradoxically, when the courtship goes well, they display aggressive
behaviour towards each other – violent bill fencing, signalling fear by lifting wings, grabbing each
other by the talons and bills, and the female chasing the male to take the prey he has brought her.

During the long pre-laying period, extending from winter to spring in Europe, females roost inside
the nest cavity more often than males. As the season progresses, females as well as males roost more
frequently in the nest cavities and become vocally more conspicuous. The male presents his nest to
the female more often and courts her by offering prey items. These offerings are essential, because
females need to store energy to produce eggs and become so heavy that they can hardly hunt. A male,
therefore, must feed his partner generously if he wants a clutch. One month before egg laying, males
bring to the nest an average of two voles per day, and five days before egg laying, when both sexes
roost inside the nesting cavity to copulate, the male brings five voles per day. In Germany, one male
brought more prey items per day during egg laying (5.7 on average), hatching (8.4) and the first five
days of rearing (15.8) than during incubation (4.8). The number of prey items brought by the male
thus coincides with food need, but the frequency of courtship feeding does not predict clutch size.

Barn owls copulate very frequently.

We could say that barn owls like making love. **Candlelit dinner** ... In a French study, just before copulating the male offered a prey item to the female in 40% of cases. Males are not always generous, because in a 24-hour period the number of copulations may exceed the number of food items brought, and furthermore, females do not always consume the prey that is offered. **Sweet words** ... Before 90% of copulation events, the female produces loud calls. **Moaning** ... Males and females call before and while copulating, and at the end of the act, the male emits an ecstatic snoring call. **Almost year-round** ... In Germany, barn owls can copulate any time from February, two months before egg laying, until the end of the breeding season in November. Not all individuals are so insatiable, and copulation usually starts 20–50 days before laying the first clutch and 17–33 days before the second clutch of the year. **Eager to copulate** ... Most copulation acts are performed at the nest as soon as the male enters the nesting cavity. He apparently cannot wait any longer. **Exhibitionist** ... Owls do not hesitate to copulate beside their offspring. **Extended physical contact** ... The duration of mounting is relatively long compared to that seen in other birds, with 29 seconds on average and an observed maximum of 1 minute – but this does not match the aquatic warbler's record of 24 minutes.

## Mate guarding
Barn owls are commonly socially monogamous, with males securing one female at a time. While males often try to copulate outside the pair bond, they do not appreciate being 'cuckolded' themselves. In owls, the male is the main food provider, and hence, to avoid wasting energy in caring for offspring sired by competitors, a male will follow his partner as closely as possible, a strategy described as 'mate guarding'. Accordingly, just before and during laying, four radio-tagged Scottish males spent 74% of the night-time close to the nest. Once the clutch was complete, and hence the risk of being cuckolded negligible, males spent only 28% of their time close to their mate.

## Frequency of copulation
Although one copulation act would be sufficient to fertilize all the eggs of a clutch, barn owls can copulate hundreds of times in a single reproductive attempt. Why? Copulating so often with the social mate ensures that the sperm of any interloper would be diluted, preventing him from fertilizing even a single egg. Twenty days before egg laying, pairs copulate at least once a day to reach a peak just before egg laying, with more than one copulation event per hour (the maximum recorded is 70 copulations in 24 hours). As soon as egg laying starts, the rate of copulating declines (to about one

copulation per hour), and during incubation the birds can still copulate more than fourteen times a day. They continue to copulate during hatching and chick rearing, but once the mother starts hunting for the offspring, copulation is likely to end unless there is going to be a second annual clutch. A male may therefore continue copulating after the first clutch is complete, which seems to motivate the female to produce the second annual clutch with him rather than with another male. Although pairs that go on to produce a large clutch copulate more often each day than those that lay a smaller clutch, food supply does not appear to affect copulation frequency. These patterns seem to be general in owls and raptors.

## Age at sexual maturity
Short-lived animals usually start to reproduce at a younger age than long-lived animals: quails can breed at 60 days of age, but bearded vultures do not reproduce successfully until they are at least eleven years old. Knowledge about the age at sexual maturity is important because it affects the trajectory of life-history traits such as ageing and survival. In the barn owl, sexual maturity is reached very early in life, usually within the first year. All Tytonidae, including the largest sooty and grass owls, can reproduce in their first year. In the USA and Europe, more than 95% of barn owls breed for the first time within their first year, and only a few individuals delay until they are two years old. The onset of reproduction can be postponed if a bird is unable to attract a partner or if environmental conditions are poor in its first year of life, as shown by a study in the Netherlands. The earliest reproduction observed in barn owls was in captivity with individuals aged only 180 days. In 283 Swiss barn owls breeding in their first year of life, the first egg was laid at 332 days of age, on average, with the youngest individual being only 207 days and the oldest 438 days. Interestingly, this does not vary between the sexes.

## FUTURE RESEARCH
- Inter-individual variation in the timing of the breeding season should be studied, to identify the characteristics that are associated with early breeding onset.
- Identify the variables (e.g. temperature, date, time of day, morphology) that can explain the variation in courtship feeding and copulation behaviour.
- Investigate whether sexual activity differs between individuals, and why.
- Inter-individual variation in the age at which owls breed for the first time should be studied, to understand which individuals are likely to breed earlier in life.

## FURTHER READING
Antor, R. J., Margalida, A., Frey, H., Heredia, R., Lorente, L. and Sesé, J. A. 2007. First breeding age in captive and wild bearded vultures *Gypaetus barbatus*. *Acta Ornithol.* **42**: 114–118.

Bruijn, O. de. 1994. Population ecology and conservation of the barn owl *Tyto alba* in farmland habitats in Liemers and Achterhoek (the Netherlands). *Ardea* **82**: 1–109.

Durant, J. M., Gendner, J.-P. and Handrich, Y. 2010. Behavioural and body mass changes before egg laying in the barn owl: cues for clutch size determination? *J. Ornithol.* **151**: 11–17.

Epple, W. 1985. Ethologische Anpassungen im Fortpflanzungssystem der Schleiereule (*Tyto alba* Scop., 1769). *Ökol. Vögel* **7**: 1–95.

Marti, C. D. 1997. Lifetime reproductive success in barn owls near the limit of the species' range. *Auk* **114**: 581–592.

Møller, A. P. 1987. Copulation behaviour in the goshawk, *Accipiter gentilis*. *Anim. Behav.* **35**: 755–763.

Muller, Y. 1981. Une chouette effraie (*Tyto alba*) se reproduit dès l'âge de 7 mois. *Ciconia* **2–3**: 143–146.

Platz, M. 1996. Untersuchungen zur Brutbiologie eines Schleiereulenpaares (*Tyto alba*) unter besonderer Berücksichtigung des Nahrungserwerbs in der Agrarlandschaft. Diplomarbeit Freie Universität/Humboldt Universität Berlin.

Roulin, A. 1996. Balz und Paarbildungserfolg bei der Schleiereule *Tyto alba*. *Ornithol. Beob.* **93**: 184–189.

Roulin, A. 1998. Formation des couples en hiver chez l'effraie des clochers *Tyto alba* en Suisse. *Nos Oiseaux* **45**: 83–89.

Schulzehagen, K., Leisler, B., Birkhead, T. R. and Dyrcz, A. 1995. Prolonged copulation, sperm reserves and sperm competition in the aquatic warbler *Acrocephalus paludicola*. *Ibis* **137**: 85–91.

Stearns, S. C. 1992. *The Evolution of Life Histories*. Oxford: Oxford University Press.

Ritual behaviour
between a male and
female barn owl.

## 7.2   MATING SYSTEM

# Frivolous behaviour

**Barn owls are generally solitary breeders, and hence they usually secure one mate at a time (social monogamy), as in most other birds. A few males can be polygynous, with their two females sometimes even nesting within the same cavity, or perhaps in nests several kilometres apart. In contrast, few females produce a brood with two males (polyandry). Grass owls, and sometimes barn owls, can be gregarious and form loose colonies where they have the opportunity to copulate with several mates. Approximately 1 in 50 nestlings is not sired by the male that feeds it, and 1% of the breeding pairs are incestuous.**

Information about how many partners animals secure (the mating system) is important when evaluating whether the descendants of one population result from the reproduction of a few or most sexually mature individuals. **Polygyny**, where one male secures multiple females, is frequent in mammals because a female does not need extra help from her mate to raise the pups, which drink maternal milk. For this reason, males compete to copulate with as many females as possible. In birds, **social monogamy** is more common than in mammals because raising offspring usually requires the two parents to search for food. In owls, the male collects most of the food for the offspring, while the female incubates the eggs, broods the young and distributes the food among them until they can eat by themselves. As observed on occasion, a male can use his free time to attract other females when his first mate is taking care of the eggs. In contrast, a female only has time to seduce other males once her offspring can maintain their own body heat and need no help to eat the food brought to the nest.

**Breeding density**  Mating systems are closely associated with social behaviour. For instance, compared to solitary breeding species, colonial and gregarious birds have more opportunities to meet potential partners and become polygamous. Although most Tytonidae breed solitarily, probably because their mammalian prey is not abundant enough to support many owls, they can sometimes form loose colonies. In the African grass owl, nests can be as little as 150 metres apart, and in Australian grass owls, 40–50 pairs have been recorded in 40-hectare fallow rice fields. Although ornithologists in Europe usually consider one pair of barn owls to breed per village, densities can easily increase if many potential nest sites are available, as illustrated by the following examples:

- In South Africa, two nests were spaced only 20 metres apart.
- In Zimbabwe, five nests were found in a 100-metre-long mine dump.
- In Mali, up to 40 pairs were recorded breeding in an area of 250 hectares.
- In France, three females were breeding in the same church.
- On the Canary Islands, three cavities contained a barn owl family within a radius of 40 metres.
- In the USA, nine nests 4.5–140 metres apart were found in a deserted petrol station.
- On the Galápagos Islands, two nests were only 4 metres apart, and a third nest was 60 metres away.
- In England, three females were seen breeding in the same farmhouse roof with two males.

In contrast, the larger Australian Tytonidae are less gregarious. In the masked owl, the shortest distance between two nests is normally 1 kilometre, and up to nine pairs of sooty owls have been found in an area of 300–700 square kilometres.

**Communal breeding**  As the above examples suggest, barn owls can be tolerant of conspecifics: two females can breed very close to each other, even inside the same nest box with a single male, as reported in the USA and Europe. The most extraordinary example comes from Israel, where a single male copulated with two females that laid their clutches close to each other. The male frequently copulated with one female while the other incubated her eggs. The two females laid a total of 20 eggs in the same nest cup, of which 19 hatched and 16 offspring fledged.

**Polygyny**  One male securing two females simultaneously in two different sites is not so rare. In Scotland, seven such cases were recorded out of 419 nesting attempts (1.7%). Polygyny occurs mainly in years in which many barn owls produce two successive annual broods, as seen in Germany.

**No jealousy.** Females can sometimes be aware of their male having a 'lover' without, apparently, interfering in the affair. This is obvious from the above example from Israel, where two females bred very close to each other: they both apparently accepted the situation without showing any signs of trying to become the male's favourite mate. Maybe they were sisters, which could explain their tolerance. However, in Switzerland a similar case of two females breeding with a single male in the same nest box has been observed. The two females were not sisters, and using molecular methods it was possible to prove that they both produced offspring successfully.

When offspring are mixed within a single communal nest, the adults are unable to determine to which mother a nestling belongs. Barn owls do not even differentiate between a kestrel nestling and their own offspring. Indeed, barn owl females have been observed to incubate the eggs of a kestrel along with their own eggs and then feed the young barn owls and kestrels until the fledging stage.

**A favourite?** No comprehensive study has determined whether polygynous males invest more effort to raise one of their two broods.

**Secret partner.** Are barn owls supporters of free love? Apparently not. On many occasions, the two females of a polygynous male laid their eggs at sites up to 4 kilometres apart, which suggests that each female may not have been aware that her male had another partner.

**Polyandry and cooperative breeding**  One female securing two males simultaneously has been observed only in captivity when two males and two females were kept in an aviary. Although

A communal barn owl nest in Israel. *Top*: The male is bringing one prey item to one of his two mates. *Middle*: The male is copulating with one of the two females. *Below*: The two females are brooding their chicks. Drawings after pictures by Ezra Hadad.

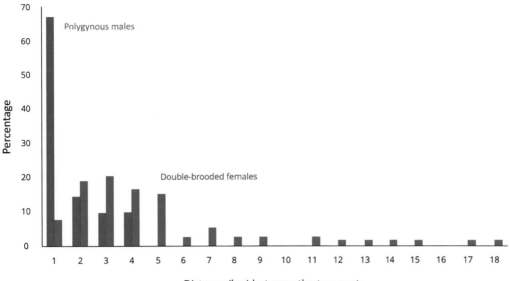

Frequency distributions of distances between the two nests of polygynous males and the two nests of females who changed mate and site to produce their second annual brood. Because the father captures most of the prey items for his family, polygynous males raise two families with different partners in sites that are relatively close to each other. Once the nestlings can eat food by themselves and can regulate their body temperature without the help of their mother, a female can desert her family and initiate a second annual breeding attempt with a new partner quite far from her first nest. The first male continues to take care of the first nestlings alone.

the two males fed both females and were not aggressive towards each other, one male copulated with both females, of which only one laid eggs, while the other male copulated mainly with just one female. In another group of four individuals, one male prevented the other male from copulating. In these two cases, only one female bred, and the two males and the other female participated in feeding that female and her offspring. Similar observations were made in captive Australian grass owls, where one female produced a clutch, and another female helped to raise the chicks. In the wild, three observations showed that a female fed the brood of another female, suggesting that cooperative breeding may occur under natural conditions as well. It is unclear whether these behaviours are common or characteristic of deficient individuals. We have even observed a kestrel feeding a brood of barn owls in Switzerland!

**Extra-pair fertilization**  In monogamous organisms, where individuals secure one mate at a time, males can increase their reproductive success by copulating with multiple partners. Females may copulate with males outside the pair bond if their social mate is not of sufficient genetic quality or not genetically compatible. In socially monogamous birds, **extra-pair offspring** have been found in approximately 90% of species, and in some species more than 11% of offspring are sired by an extra-pair male. However, as in other owls and raptors, barn owls copulate so frequently inside the pair bond that the probability of another male fertilizing an egg is low. The female sexual tract contains so much sperm from her social mate that the amount of sperm transferred by a competitor is like a drop of water in the ocean. Accordingly, in Switzerland, only 27 out of 1403 nestlings (1.9%) were sired by males other than the one taking care of them. Copulating outside the pair bond is therefore not an efficient strategy for increasing fitness in the barn owl.

**Incest**  Breeding with a closely related mate can be detrimental for the offspring. Incest leads to consanguinity, i.e., genetic disorders resulting from homozygosity; consanguinity increases the

Copulating inside the nest cavity.

risk of developing detrimental traits encoded by rare recessive alleles. If one family possesses such recessive alleles, individuals may not be affected as long as they all have a copy of the dominant non-detrimental allele. However, once these related individuals breed together, the likelihood that they give birth to offspring possessing two copies of the recessive allele becomes non-negligible.

As in most birds, in which inbreeding accounts for only 0–6% of all mating, the probability of pairing with a related individual is low in the barn owl. In Scotland and Switzerland, only one of 157 pairs (0.6%) and ten out of 830 pairs (1.2%) were composed of closely related partners, respectively. In the USA and Europe, eight breeding pairs from siblings born in the same nest were recorded, along with one brother–sister pair (born in different years), four mother–son pairs, two father–daughter pairs, one grandmother–grandson pair and one aunt–nephew pair. These inbred pairs were located at a distance of a few kilometres from the birth site, emphasizing that the best way to avoid breeding with a closely related individual is to disperse far from the natal site. Recognizing family members to avoid breeding with relatives may be another solution, although nothing is known about such a mechanism in Tytonidae.

## FUTURE RESEARCH

- The frequency of polygyny should be assessed by tagging males with GPS. This would offer the possibility of checking whether males feed one or two families.
- Once polygynous males are detected, genetic analyses (or ringing) could help determine whether the two females of these males are closely related.
- The reproductive success of polygynous males should be recorded in their two nests to examine whether one family is preferentially fed over the other. Of further interest would be identifying why the polygynous male may prefer one family over the other.
- In small populations, such as those living on islands, the level of inbreeding could be high because the probability of encountering close relatives may be greater. The degree of inbreeding should therefore be monitored in small and large populations, with important implications for conservation.
- The level of extra-pair paternity, although generally low in the barn owl, might be higher in dense populations where the opportunity to mate outside the pair bond is high. It would therefore be interesting to investigate the possibility of extra-pair paternity in populations differing in density.
- Although the level of extra-pair paternity is low, nothing is known about the rate of extra-pair copulation. Behavioural studies should be performed to examine whether barn owls must copulate outside the pair bond very frequently to sire at least one extra-pair offspring.
- To avoid inbreeding, owls may recognize relatives and avoid breeding with them. Behavioural experiments should be carried out to test whether owls are able to recognize relatives.

## FURTHER READING

Ducret, V., Gahier, A., Simon, C., Goudet, J. and Roulin, A. 2016. Sex-specific allelic transmission suggests sexual conflict at *MC1R*. *Mol. Ecol.* **25**: 4551–4563.

Dunlop, R. and Pain, P. 2016. Reproduction, social behavior and captive husbandry in the eastern grass owl (*Tyto longimembris*). *J. Zoo Aquar. Res.* **4**: 169–173.

Epple, W. 1985. Ethologische Anpassungen im Fortpflanzungssystem der Schleiereule (*Tyto alba* Scop., 1769). *Ökol. Vögel* **7**: 1–95.

Greenwood, P. J. 1980. Mating systems, philopatry and dispersal in birds and mammals. *Anim. Behav.* **28**: 1140–1162.

Griffith, S. C., Owens, I. P. F. and Thuman, K. A. 2002. Extra pair paternity in birds: a review of interspecific variation and adaptive function. *Mol. Ecol.* **11**: 2195–2212.

Hadad, E., Roulin, A. and Charter, M. 2015. A record of communal nesting in the barn owl (*Tyto alba*). *Wilson J. Ornithol.* **127**: 114–119.

Marti, C. D. 1990. Same-nest polygyny in the barn owl. *Condor* **92**: 261–263.

Ralls, K., Harvey, P. H. and Lyles, A. M. 1986. Inbreeding in natural populations of birds and mammals. In Soulé, M. E. (ed) *Conservation Biology: the Science of Scarcity and Diversity*. Sunderland, MA: Sinauer Associates, pp. 35–56.

Roulin, A., Müller, W., Sasvári, L., Dijkstra, C., Ducrest, A.-L., Riols, C., Wink, M. and Lubjuhn, T. 2004. Extra-pair paternity, testes size and testosterone level in relation to colour polymorphism in the barn owl *Tyto alba*. *J. Avian Biol.* **35**: 492–500.

Barn owls can be very attentive to their mate.

## 7.3  FIDELITY AND DIVORCE

# Eternal love or leaving home

**Fidelity and divorce are the two sides of the same coin for achieving high reproductive success. Barn owls search for a soul mate, a process that takes time; this might explain why yearling males are more likely than older males to separate from their partner. When reproductive success is low, there is a strong temptation to blame the partner and divorce to restore reproductive success with another mate. On average 20% of adults divorce, but once an appropriate partner has been found, barn owls remain faithful, which improves coordination in reproductive activities over time.**

Albatrosses pair for life, and house martins break up after each breeding attempt. There is thus a significant variation in divorce rates among birds, which requires an explanation. Animals benefit from familiarity with the area where they live, and long-term partnership helps improve coordination in reproductive activities. When a partner is not of high enough quality, or if the combination of paternally and maternally inherited genes is detrimental to the offspring, reducing their survival prospects, a solution is to divorce and find a new partner. This solution may also imply changing nest site as well, particularly for females, who in most species are less territorial than males.

**Nest-site fidelity** Even if owls occupy the same site over many years, these sites are not always occupied by the same individuals. Farmers are always surprised to learn that year after year different barn

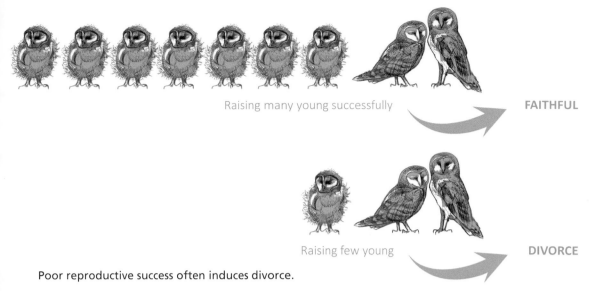

Raising many young successfully → FAITHFUL

Raising few young → DIVORCE

Poor reproductive success often induces divorce.

owls breed in the nest box on their farm. Over a period of eight years, six different females bred in one French nest box. In New Jersey (USA), from one year to the next, a different set of breeding females and males occupied the same nest boxes in 65% and 33% of cases, respectively. This high turnover of breeding birds increases with mortality and immigration, as well as with the abundance of nest sites. Indeed, the more sites are available, the more opportunities are offered to resident birds to move to a nearby site. In regions where cavities are rare, birds may have no other option than to stay at the same site. This might

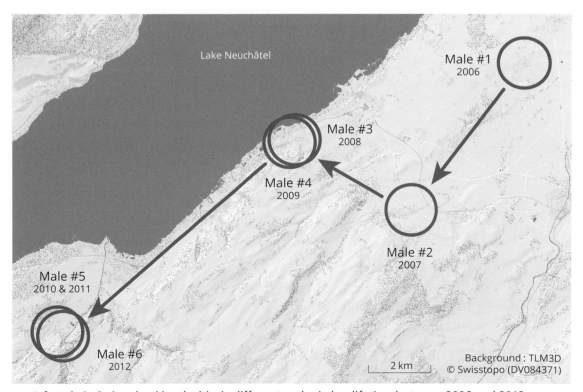

A female in Switzerland bred with six different males in her lifetime between 2006 and 2012.

be the reason why in Scotland, 99% of the breeding males and 95% of the females were faithful to their nest site, and why in Germany, females were found to breed in the same nest box up to seven years in a row.

Males typically defend a nest site to attract females, who visit several potential partners before making the decision to settle. The site-faithful males can divorce without moving to a new site, as they just need to expel their mate out of their territory and wait for a new female to visit them. In contrast, females must leave their home to break up the pair. However, in birds, as in most other animals, females are more careful than males in their decision to mate with a given mate. Thus, females initiate divorce more often than males, something that has not yet been confirmed in Tytonidae. In Switzerland, only 40% of the divorced males moved to another site, which on average was only 1.1 kilometres from the previous site. This indicates that males defend a territory including several potential nest sites. In contrast, 96% of divorced females, compared with 22% of faithful females, produced a subsequent brood in a site that was relatively far from the previous site at a mean distance of 2.3 kilometres. This indicates that females are less faithful to the breeding site, and if they want to breed with another male they must change territory. Similar behavioural patterns were observed in Germany.

**Mate change** Animals can change mates either because they divorce or because their partner dies. The pair bond is not so strong. Over a period of three years, a German female barn owl

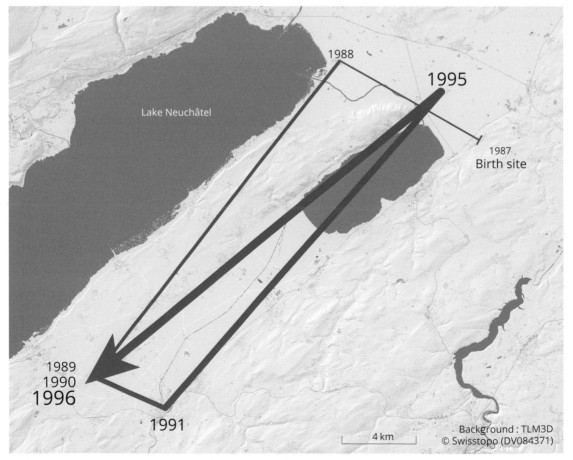

A female in Switzerland bred in four different nest sites in her lifetime. Born in 1987, this female bred fairly close to her birth site the year after. Then she moved a relatively long distance to breed in two nest boxes in three consecutive years (1989, 1990 and 1991), before returning close to the birth site in 1995 and finally returning in 1996 to the same nest box as in 1989 and 1990.

Should I stay or should I go? The eternal question, even for the barn owl (*Tyto alba sertens*, India).
© Jim Parkås.

produced five successful broods, each time with a different partner and in a different nest box. In a Swiss population, 60% of the males and 50% of the females changed mates between successive years. Out of 758 different pairs, 634 (83.6%) bred together for only one year, 82 (10.8%) for two years, 27 (3.6%) for three years, 12 (1.6%) for four years, two (0.3%) for five years and one (0.1%) for six years. Divorce between two successive years occurs relatively frequently, with 23.5% and 18.5% of the pairs divorcing in Switzerland and Germany, respectively.

**Causes of divorce** Divorce is defined as when an individual changes mates while its previous partner is still alive. The primary function of divorce is to secure a higher-quality or more compatible partner than the previous mate. This aspect was studied in a Swiss population over 26 years. Two factors were associated with divorce, poor reproductive success and male experience:

- **Poor reproductive success.** Divorce was particularly frequent after the complete failure of a reproductive attempt. Ten percent of the pairs that divorced had not produced any fledgling the year before, while only 2% of the pairs remaining faithful had failed in their breeding attempt. Among the pairs that produced at least one fledgling, those who fledged more offspring (on average 4.7 fledglings) were less likely to divorce the following year than were pairs who fledged fewer offspring (on average 3.5 fledglings).
- **Experience.** Finding a soul mate is a long process that requires sampling several potential mates before deciding which partner is the most suitable. This is particularly true in males. Pairs were more likely to divorce if the male was a yearling rather than older, while female age had no impact on divorce.

**Benefits and costs of divorce** As shown in Switzerland, divorce is beneficial because it allows unfortunate individuals to restore reproductive success. Newly paired divorcees produced more offspring than they did the year before divorcing, and therefore they raised a similar number of fledglings as faithful pairs. However, divorce is not cost-free. Acquiring a new mate can take time. Divorcees produce a clutch with the new partner later in the season than do individuals who remain faithful to their partners. Individuals who divorce may have few options to select a new mate simply because the best potential mates are faithful to their high-quality partners. The new females of divorced males had smaller black spots on the tips of their ventral feathers, a signal of quality, but were younger and thus less experienced than the females with whom those males bred the year before divorcing. Finally, divorcees often must leave their nest site to find a new mate and are therefore relegated to poorer-quality sites.

**Benefits of long-term monogamy** While divorce may be the best option when a barn owl is paired with a poor-quality or a non-compatible mate, long-term monogamy improves the efficiency of coordination between partners in reproductive activities. The probability that a hatchling fledged successfully increased with the number of years that the parent owls bred with the same partner. Faithful partners became so efficient that they were even able to reduce clutch size to produce the same number of fledglings. Therefore, barn owls look for a long-term relationship to improve coordination, and to reach this goal they sometimes should divorce if they are not paired with the most suitable mate.

## FUTURE RESEARCH
- The relative importance of immigration and availability of nest sites in determining whether an individual will divorce or change site should be studied.
- Whether males or females initiate divorce should be investigated.

## FURTHER READING
Choudhury, S. 1995. Divorce in birds: a review of the hypotheses. *Anim. Behav.* **50**: 413–429.
Dreiss, A. N. and Roulin, A. 2014. Divorce in the barn owl: securing a compatible or better mate entails the cost of re-pairing with a less ornamented female mate. *J. Evol. Biol.* **27**: 1114–1124.
Kniprath, E. 2011. Scheidung und Partnertreue bei der Schleiereule *Tyto alba. Eulen-Rundblick* **61**: 76–86.

# 8 Reproduction

8.1   Nest sites
8.2   Interspecific competition over nest sites
8.3   Reproductive season
8.4   Egg formation
8.5   Clutch size
8.6   Incubation
8.7   Hatching
8.8   Brood size
8.9   Nestling growth
8.10  Second and third annual broods
8.11  Offspring desertion

A barn owl at the entrance to its nest, located inside a house.

# Bedroom

**The barn owl species complex (*Tyto alba*, *T. furcata* and *T. javanica*) is most famous for its propensity to breed in man-made structures, although it naturally lays eggs in cavities in trees, cliffs, caves and river banks, and even underground. This flexibility to breed in so many sites may fuel the barn owl's successful colonization of the world. Other Tytonidae, which are more specialized in their nest-site choice, are much less successful. For example, grass owls breed exclusively on the ground in grass-lands, and masked owls and sooty owls breed primarily in tree cavities, rarely inside caves, and very rarely inside man-made structures.**

Knowledge of where Tytonidae breed is important to target conservation measures protecting crucial habitat elements for these species. Owls, like many other animals, are currently facing substantial pressures from habitat loss and disturbance. Owl species that breed exclusively in one type of site are vulnerable if these sites decline because of anthropogenic habitat alteration. For instance, forest man-agement does not favour the preservation of old trees with cavities, which are so crucial in providing nesting habitat for many animal species. Furthermore, at least in Europe, most buildings are now more hermetically sealed, preventing owls from finding sites in which to lay eggs.

Although the barn owl's habit of breeding in human constructions is widespread, there are still pronounced regional differences. For instance, North American barn owls breed more often in natural cavities, such as inside trees, than European barn owls. In North America, vast areas free of man-made structures may force barn owls to find other nest sites. Where buildings are rare, this species indeed

Barn owls and their relatives can use a range of nest sites. *Left*: A grass owl breeding on the ground in Taiwan. © Yi-Shuo Tseng. *Right*: A barn owl breeding in a tree cavity on the island of Sumba, Indonesia. © Thierry Quelennec

regularly roosts and breeds in tree holes, as observed on islands in Southeast Asia, the Pacific, São Tomé and Mayotte. In regions without nest boxes, such as in New Jersey (USA) or Pakistan, three-quarters of pairs breed in trees and a quarter in man-made structures. In Europe, breeding in tree holes is rare on the mainland but frequent in the United Kingdom. Where both human constructions and trees are rare, barn owls breed in natural cavities in cliffs, in caves, in the walls of lava tubes and even in the ground – as observed in Madeira, Tenerife, Sardinia, Cap Verde, the Middle East and Galápagos.

Across most of the world, the use of other birds' nests is rare. In North America, for example, there have been just a few recorded cases of barn owls breeding in old nests of ospreys, woodpigeons and crows. In Africa, however, barn owls commonly take over the domed nests of hamerkops and sociable weavers. In Mali, every one of 178 detected barn owl clutches was in a hamerkop nest, while 53 of 107 clutches in South Africa and 11 of 28 clutches in Zimbabwe were found in hamerkop nests.

## Why do barn owls breed in man-made structures?

Like house sparrows and feral pigeons, the barn owl has adapted well to human development. Before people started providing them with nest sites, most barn owls probably bred in tree cavities and in cliffs up to 450 metres above the ground, as seen in Scotland. Houses, barns and temples resemble cliffs, but why do owls generally prefer to breed in man-made structures rather than in natural sites, such as in cliffs or trees? Several non-mutually exclusive explanations are plausible:

- **Thermoregulatory protection.** Barn owls may breed (and roost) in man-made structures because these sites are more insulated against cold or warm temperatures. Accordingly, in the United Kingdom, where the climate is milder during the winter and early breeding season than on the continent, owls nest more often in trees. In California, ambient temperatures can be high, and breeding in buildings, just under the roof, may be risky as these locations are prone to extreme highs. This could partly explain why barn owls bred in man-made sites only 28% of the time in California, preferring natural sites.
- **Protection against precipitation.** In a study in the United Kingdom, 90% of barn owls bred in buildings in regions where rainfall exceeds 1000 millimetres per year, whereas where it rains less than 550 millimetres a year, 70% of barn owls nested in trees. Indeed, water drips down tree trunks and can accumulate inside nest cavities, where the eggs and chicks can be flooded.
- **Man-made sites are more stable across years.** Cavities are usually a limiting resource, and competition to take over such sites can be intense. Because natural sites, such as those in old trees,

06:19:03

06:19:12

06:20:11

06:20:22

A barn owl is breeding in a nest box fixed inside a barn, just behind the wall. Although it is still dark, a European kestrel visits the nest box less than a minute after the barn owl enters the box. Photos taken with a camera trap in Switzerland. © Robin Séchaud & Kim Schalcher

are frequently destroyed by high winds or disintegrate after heavy rains, individuals breeding in trees may experience more stress to find a new breeding site than owls breeding in man-made structures. For instance, in the USA and Canada, approximately 35% of trees with known barn owl nests were destroyed in four years, while during the same period only 6% of man-made nesting structures were lost. A single storm in Britain and Ireland destroyed 19% of barn owl tree nests.

- **Protection against predators.** Possibly, fewer predators and passerines that mob owls may occupy man-made structures. However, given that these owls may suffer from human disturbance, is the overall level of disturbance globally lower in nests located in man-made structures or in natural cavities?
- **Reduced interspecific competition for nest sites.** Competition for tree cavities may be more intense than for nest sites located in farm buildings because more species breed and shelter in trees than in man-made structures. For this reason, barn owls may have found a way to escape interspecific competition by breeding closer to humans. Many bird species breed in cliffs, yet the barn owl is one of the few birds that have adapted to colonizing man-made structures.
- **Close to foraging sites.** Often, suitable human structures are built close to the agricultural lands that barn owls are well adapted to hunt in. Owls that breed inside man-made structures may therefore be closer to their foraging sites than owls breeding in other sites such as cliffs.

If many birds breed in cliffs, why has the barn owl succeeded in colonizing man-made structures to breed and roost? Are barn owls particularly resilient to human disturbance? Their medium size and nocturnal habits may prevent barn owls from being easily detected by humans. Furthermore, barn owls may be more resistant than other birds to stress and thus better able to cope with the presence of humans. Even if this might be the case, human disturbance can still entail costs. A study in Switzerland showed that barn owls are physiologically stressed by the presence of humans, and in Canada, Israel, Portugal, Spain, Switzerland and the United Kingdom, sites surrounded by more roads (and with high traffic volume) are occupied less often than sites located far from roads. Barn owls are attracted by sites offered by humans, but not by humans themselves!

**Preference for nest boxes**    Throughout most developed parts of the world (except in Australia), barn owls breed in man-made structures, from churches, temples and minarets to barns, farmhouses and ruins – and especially in nest boxes. Boxes fixed against barns, against trees or on poles in the middle of agricultural fields are often so successfully adopted by barn owls that they prefer them over traditional breeding sites, which are consequently abandoned. In the Czech Republic, the proportion of owls breeding in churches decreased from 51% in 1940 to 2% in 2007, and in one German population the percentage of pairs breeding in nest boxes increased from 17% in 1977 to 100% in 1987. Similar phenomena have been observed in other European countries, probably because people are more inclined to prevent birds from having access to church towers and because barn owls appreciate well-designed nest boxes offering prime breeding conditions. The preference for nest boxes can be explained by the fact that reproduction is often more successful in boxes than in other man-made sites, as observed in the USA, Argentina and the Netherlands. Nest boxes should be fixed in places safe from predators, high above the ground (usually above 5 metres), in the shade and not exposed to rain and wind.

**Nest sites of other Tytonidae**    If barn owls are generalists in their nest-site choice, other Tytonidae are rather specialists. African and eastern grass owls breed exclusively on the ground in dense grass away from trees. Sooty owls breed in large tree hollows up to 40 metres above the ground and always in dense forests. Australian and Tasmanian masked owls breed in eucalypt trees in open woodlands, closed forests and even isolated trees in agricultural landscapes. Data on other Tytonidae are disparate. Minahassa masked owls live in dense montane forests, and one nest has been found in a tree hole. Two nests of the Sulawesi masked owl were discovered in tree holes not far from a village, and another nest was found inside a cave 10 metres above ground. Finally, the very rare red owl breeds in tree holes in the dense evergreen forests of Madagascar.

**Nest construction**    As in other owl species, the preparation of a nest is rudimentary in Tytonidae. Although the parents do not bring any material to build or consolidate their nest, the mother can break up regurgitated pellets to form a small nest depression where she lays the eggs. However, in newly used nest sites, pellets are rare; therefore, ornithologists often add a suitable substrate inside the nest box. In regions where cavities are rare or where ambient temperatures are high, owls may work hard to construct a refuge. Barn owls can excavate a long tunnel (up to 132 centimetres in length) by scratching the soil with their talons in vertical stream banks and arroyos, a behaviour that is relatively frequent in North America.

## FUTURE RESEARCH

- An understanding of why barn owls breed in tree holes more frequently in some regions than in others is still required. Analyses should be performed to test whether the propensity to use different tree types is related to climatic variables or to human activities.
- Are the negative impacts of disturbance by humans, predators, competitors and other animals less pronounced in breeding sites located in man-made structures than in natural sites?

## FURTHER READING

Almasi, B., Béziers, P., Roulin, A. and Jenni, L. 2015. Agricultural land-use and human presence around breeding sites increase stress-hormone levels and decrease body mass in barn owl nestlings. *Oecologia* **179**: 89–101.
Blaker, G. B. 1933. The barn owl in England. *Bird Notes and News* **15**: 207–211.
Shawyer, C. R. 1987. *The Barn Owl in the British Isles: its Past, Present and Future.* London: Hawk Trust.

Ruddy shelducks can breed in the nest boxes specifically prepared and installed for barn owls. Photo taken with a camera trap in Switzerland. © Robin Séchaud & Kim Schalcher

## 8.2   INTERSPECIFIC COMPETITION OVER NEST SITES

# Disapproval, tolerance and indifference

**Animals compete intensely to protect their stake in a nest site, a resource that is often scarce. Competing with other owls, raptors and other predators to secure a nest site can be dangerous and time-consuming, and may have high energy costs. This may explain why barn owls often tolerate the presence of other animals breeding close to them.**

Once a barn owl acquires a cavity, it denies access to others unless the site is large enough to share. Tolerating the presence of other animals has costs: it can increase conspicuousness to predators, favour the spread of parasites, generate interference, and disturb resting and sleeping activities. Collective defence of the cavity may compensate for these drawbacks, and barn owls may also benefit from predators with which they share a cavity, as they scare or attack conspecifics or other animals. However, so far, no clear benefits of breeding close to other birds have been reported in Tytonidae.

The size of the entrance to a cavity determines who can occupy it. Animals smaller than barn owls, such as scops owls, jackdaws, house sparrows and rats, are not powerful enough to compete with a barn owl, and they can occupy large nest boxes only if the entrance is small enough to prevent owls from entering, as observed in Israel. In contrast, the kestrel, with its sharp bill and claws, can injure barn owls and will sometimes usurp a cavity from an owl. However, the usual situation is that barn owls

Several birds, even those that do not breed in cavities, such as the carrion crow (*left*), may be curious enough to visit barn owl nest boxes. Martens (*right*) visit the boxes to consume barn owl eggs and chicks. Photos taken with a camera trap in Switzerland. © Robin Séchaud & Kim Schalcher

outcompete kestrels. Being attacked at night by owls is stressful for diurnal falcons, which can see almost nothing in the dark. In Israel, 22% of kestrels abandoned the clutch that they had laid in a nest box because of owl disturbance.

While these observations may suggest that the relationship between kestrels and barn owls is always tense, these birds can coexist. In France, Germany, Israel, Switzerland and the United Kingdom, kestrel and barn owl pairs have been recorded breeding simultaneously and successfully in the same nest box. Sometimes, the kestrel lays its eggs close to the nest entrance in a narrow corridor, and at night, when the barn owls leave and return to the nest, they walk on the kestrel's back. Successful sharing of a nest cavity has also been observed with a tawny owl pair, feral pigeons, jackdaws and even hornets.

Although barn owls and their relatives are equipped with sharp claws and bills, individuals may choose to avoid costly conflicts. Although eagle owls can eat barn owls, these species sometimes breed a few metres from each other – perhaps because in some cases barn owls are attracted by the protection against other predators that eagle owls can confer. Grass owls are also very tolerant of other species, and can breed within 20–50 metres of marsh owls and African marsh harriers.

## FUTURE RESEARCH
- The costs and benefits of breeding close to other birds should be identified to understand why barn owls either compete to secure a cavity or tolerate the presence of conspecifics or other animals.
- Barn owls that share a cavity with other animals may be less bold than owls that compete fiercely to monopolize a nest site. Barn owl personality could be assessed in the field and related to the degree to which they tolerate the presence of other animals.

## FURTHER READING
Charter, M., Izhaki, I. and Leshem, Y. 2010. Effects of the risk of competition and predation on large secondary cavity breeders. *J. Ornithol.* **151**: 791–795.
Hasenclever, H. and Tiemeyer, V. 1991. Brutgemeinschaft des Turmfalken (*Falco tinnunculus*) und der Schleiereule (*Tyto alba*) in einem Brutkasten. *Charadrius* **27**: 14–18.
Kniprath, E. 2003. Breeding site competition between barn owl (*Tyto alba*) and kestrel (*Falco tinnunculus*): a product of bird protection? *Eulen-Rundblick* **51/52**: 15–17.

A communal family of sixteen nestlings produced by one male and two females in Israel. © Ezra Hadad

## 8.3 REPRODUCTIVE SEASON

# Breeding like rabbits

**Deciding when to reproduce is essential to successfully raise a family. Barn owls and sooty owls, and to a lesser extent masked owls and grass owls, are very flexible in the timing of reproduction. By being able to breed in any month of the year, an individual can finely adjust when it invests in parental care in relation to local environmental conditions, health and personal reproductive history.**

In temperate regions, ambient temperatures and food supply vary strongly between the four seasons, while organisms in the equatorial and tropical regions often experience wet and dry seasons that affect food supply. Although reproduction is usually highly seasonal, species able to reproduce throughout the year may derive advantages in terms of reduced competition with conspecifics, and furthermore may be able to raise multiple families each year. For instance, in central Europe the barn owl, which exploits open landscapes, has a reproductive period that is twice as long as that of the tawny owl, which inhabits woodland. The longer breeding season gives barn owls the opportunity to produce two broods per year and to replace a clutch that fails, which is rare in the tawny owl. Knowledge of reproductive phenology is a good proxy of how flexible animals are in deciding when and how frequently they breed.

**Timing of reproduction in the barn owl**  This group of species can reproduce throughout the year in regions where the climate is clement, such as in Australasia and the Americas. Even in temperate regions, where winters can be harsh, clutches and chicks can still be found in winter, as observed in North America (British Columbia, California, North and South Carolina, Delaware, Kentucky, Illinois, Ohio, Utah, Texas, Washington DC), Europe (Croatia, Czech Republic, France, Germany, Sardinia, Switzerland) and Argentina. Although winter reproduction is very rare, these cases demonstrate the barn owl's ability to reproduce under most environmental conditions, which is not the case in all owls having similar ecology.

This high flexibility of course has a limit. In countries where winters are extremely harsh, such as Sweden or most parts of Canada, or where summer temperatures can be very high, such as Israel, barn owls reduce their breeding period to match favourable environmental conditions. For the same reason, barn owls reproduce during two major periods in Pakistan, South Africa and Tonga, hence avoiding hot weather or heavy rains and taking advantage of the highest availability of food.

**Determinants of the barn owl's laying period** Even if food supply and weather conditions are major factors accounting for inter-annual variation in laying dates, their effects differ between populations. After snowy winters, barn owls in Utah (USA) lay eggs later in the season. In France, the United Kingdom and Utah, owls initiate reproductive activities in early spring after warm winters but later in the season after colder winters. Such adjustments were not observed in Switzerland, where the snow and winter temperatures did not influence the timing of breeding season.

These regional differences in the impact of climate may be due to the relative effects of other superposed factors, such as food supply, which may allow owls to breed earlier in years when small mammals are more abundant. Accordingly, under natural conditions, Scottish barn owls laid their eggs on 23 May on average in years with a low abundance of voles and on 10 April in years of high abundance. This was demonstrated experimentally in Germany and the United Kingdom, where barn owls laid eggs one

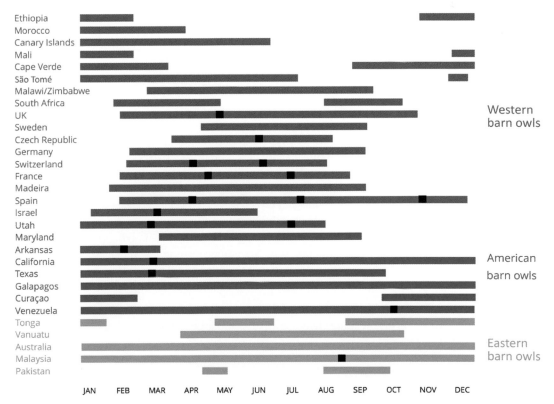

Range of breeding dates for the western barn owl (*Tyto alba*), American barn owl (*T. furcata*) and eastern barn owl (*T. javanica*). Black boxes indicate the main laying period (or periods, if the dates of more than one period are known). The absence of black boxes indicates that the main laying period has not been described in the literature. For instance, Malaysian barn owls can breed at any time of the year, with a peak of laying in August, whereas owls from Tonga show two distinct breeding seasons.

A female incubating her eggs.

month earlier when provided with extra food. However, which mechanism triggers barn owls to start egg laying on a given date? The exact date when the eggs will be laid seems to be determined about 18 days in advance, when the frequencies of visiting the nest site and of copulation increase, as seen in France.

**Timing of reproduction in other Tytonidae**  Unfortunately, little is known about the determinants of laying dates in other Tytonidae. Only data on reproductive phenology have been published; other relevant information, such as what determines phenology, is lacking. In the Australian sooty owl, there seems to be a slight peak in fledging in spring, probably due to the availability of prey species that breed seasonally. In Australia, masked owls lay eggs between February and early October, with a peak between March and July, while a peak is observed in October in Tasmania. Australian grass owls can breed during all the months of the year, although the main laying period is between March and June, while African grass owls lay from November to June. In Madagascar, the red owl starts to reproduce at the end of the rainy season so that the offspring fledge during the dry season.

## FUTURE RESEARCH
• More data are required on the breeding phenology in Tytonidae outside the barn owl species complex.

## FURTHER READING
Baudvin, H. 1986. La reproduction de la chouette effraie (*Tyto alba*). *Le Jean le Blanc* **25**: 1–125.
Chausson, A., Henry, I., Ducret, B., Almasi, B. and Roulin, A. 2014. Tawny owl *Strix aluco* as a bioindicator of barn owl *Tyto alba* breeding and the effect of winter severity on barn owl reproduction. *Ibis* **156**: 433–441.
Durant, J. M., Gendner, J.-P. and Handrich, Y. 2010. Behavioural and body mass changes before egg laying in the barn owl: cues for clutch size determination. *J. Ornithol.* **151**: 11–17.
Shaw, G. 1994. A ten year study of barn owl conservation in conifer forest. *Scot. Birds* **17**: 187–191.

Although barn owl eggs are white, they soon become brownish, particularly when the nest cavity is humid.

## 8.4  EGG FORMATION

# Long-distance runners

**Barn owls produce small eggs, which suggests that the initial investment in each offspring is relatively low. The advantage is that barn owls can afford to lay large clutches, which allows for subsequent brood-size reductions if food supply is poor. The small initial size of barn owl nestlings explains, at least in part, why they grow so slowly.**

The larger an egg is, the more resources it contains. Thus, hatching out of a large egg can confer an advantage during the first days of life. This initial advantage rapidly vanishes, however, as the amount of food delivered by the parents quickly becomes the most important factor for nestling growth and survival. Knowledge of egg size provides information about how parents invest resources in their offspring, from conception to independence.

**Pre-laying period** In contrast to 'capital breeders' such as snow geese, which accumulate resources over a long period of time to produce their eggs, the barn owl is an 'income breeder', covering the energetic needs of reproduction just before laying. This explains why food-depleted barn owls can quickly recover and lay a similar number of eggs and at a similar date as individuals who did not suffer from food depletion. In income breeders, egg production is sensitive to the environmental conditions prevailing at the time of egg laying rather than to conditions experienced long before reproduction. Even if the barn owl suffers from harsh environmental conditions, such as those met during snowy winters, this bird has the capacity to quickly catch up once conditions improve. This ability may allow barn owls to be successful income breeders.

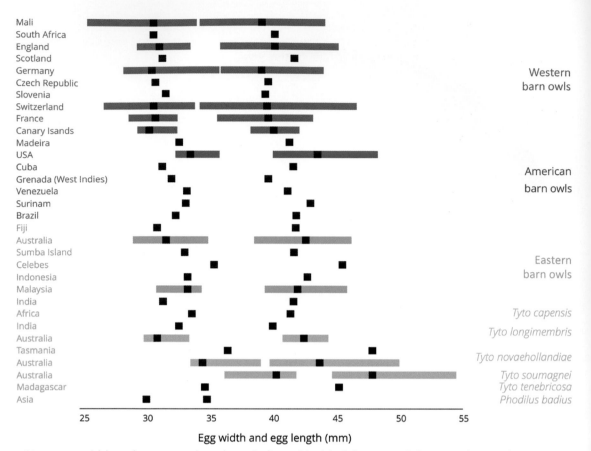

Mean egg width and mean egg length are indicated by black boxes, and the range in egg sizes by horizontal coloured bars in the barn owl species complex (western, American and eastern barn owls), grass owls (*Tyto capensis* and *T. longimembris*), Australian masked owl (*T. novaehollandiae*), red owl (*T. soumagnei*), sooty owl (*T. tenebricosa*) and bay owl (*Phodilus badius*).

How is it possible to gather the necessary resources for egg production just before laying, particularly knowing that females do not increase food intake substantially between the pre-laying and laying periods, as has been shown in France? There is a rapid increase in female body mass before the laying period, but this results largely from the accumulation of water inside tissues due to protein metabolism, and not from the storage of proteins and lipids derived from the food. Laying females accumulate 1.27 times more water than non-laying females, as well as 1.19 times more minerals and 1.11 times more proteins, but the lipid content of the laying female's body is only 63% of that of the non-laying female. The pronounced hydration of breeding females allows them to produce eggs, which are mainly made up of water (77%), followed by proteins (9%, of which 70% are stored in the albumin), minerals (8%, of which 91% are in the shell) and lipids (6%, mainly present in the yolk).

Egg production or vitellogenesis, i.e., the deposition of nutrients to form yolk in the oocyte, is rapid in birds compared to other oviparous vertebrates, in which it takes usually between one and two months. Pigeons require 5–8 days to produce eggs weighing 18 grams, kestrels need 9 days for 20-gram eggs, and shags need 13.5 days for 40-gram eggs. At least in France, barn owls need 14 days to produce eggs weighing approximately 17 grams, indicating that, compared to other birds, the rate of yolk deposition is relatively slow in the barn owl. This is because females deposit yolk in several eggs at the same time. Because approximately six eggs are developing simultaneously inside the ovary, the highest energetic demand to produce eggs occurs when the first six eggs are forming, i.e., from 1.5 days before the first egg is laid to the time of the third laying.

**Egg size**  Each barn owl egg represents 4.9% of the female's weight, which is small in comparison to other owl species of a similar size. Given the size of the bird, barn owl eggs should weigh 25 grams rather than their actual weight of about 17 grams. Barn owl eggs also contain less energy than those of kestrels or long-eared owls, which could be an adaptation for laying large clutches. A potential drawback, however, is the small chick size at birth and, in turn, the low nestling growth rate.

The production of small but numerous eggs indicates that the initial female investment in each offspring is not substantial and is compensated for by a long rearing period. The reproductive strategy of the Tytonidae is thus to spread parental investment over a long period instead of working very hard over a shorter time span. In comparison, tawny owl nestlings, which are larger than barn owl nestlings both at hatching and at fledging, have a rearing period of only one month, compared to two months in Tytonidae. When it comes to rearing their young, barn owls can thus be thought of not as sprinters but as long-distance runners. This may explain why the barn owl is so prolific, since it can produce several large broods per year. The reproductive season is therefore very long, implying that barn owls have little time to recover from reproductive activities. The low cost of producing hatchlings may also explain why parents appear to be so ready to sacrifice them, such as by letting them starve when food availability is low.

As shown in Switzerland, egg volume is less variable than clutch size and strongly depends on female identity (80% of the variance), with some barn owls laying consistently small eggs and others larger eggs throughout their life. This suggests that egg size has a strong genetic basis and depends on female size, while the environment has a weak but still significant influence. In Switzerland, the Czech Republic and Mali, eggs decrease in length as the breeding season progresses (in Switzerland egg length is, on average, 39.7 millimetres on 1 March and 39.0 millimetres on 1 August), and in both Switzerland and the Czech Republic eggs are, on average, slightly wider in large than small clutches (30.7 millimetres in ten-egg clutches and 30.4 millimetres in two-egg clutches).

## FUTURE RESEARCH
- Egg size should be measured in relation to laying sequence, to test whether females invest more resources in the first- or last-laid eggs.
- Egg size should be measured in populations around the world to examine geographic variation. This could be done both in wild populations and in natural history museums that possess egg collections.

## FURTHER READING
Altwegg, R., Schaub, M. and Roulin, A. 2007. Age-specific fitness components and their temporal variation in the barn owl. *Am. Nat.* **169**: 47–61.

Chausson, A., Henry, I., Almasi, B. and Roulin, A. 2014. Barn owl (*Tyto alba*) breeding biology in relation to breeding season climate. *J. Ornithol.* **155**: 273–281.

Durant, J. M., Massemin, S., Thouzeau, C. and Handrich, Y. 2000. Body reserves and nutritional needs during laying preparation in barn owls. *J. Comp. Physiol. B* **170**: 253–260.

Durant, J. M., Massemin, S. and Handrich, Y. 2004. More eggs the better: egg formation in captive barn owls (*Tyto alba*). *Auk* **121**: 103–109.

Meijer, T. and Drent, R. 1999. Re-examination of the capital and income dichotomy in breeding birds. *Ibis* **141**: 399–414.

A barn owl clutch of eight eggs.

## 8.5    CLUTCH SIZE

# Prolific laying

**Food supply and climatic conditions strongly impact the number of eggs that female barn owls lay. This contrasts with egg size, which is mainly sensitive to genetic factors. Current environmental changes affecting habitat quality could explain why clutch size in Europe has increased over the past fifty years.**

Compared to owls of a similar size, barn owls are very prolific and lay large clutches. Masked owls, sooty owls, bay owls and, to a lesser extent, grass owls lay relatively small clutches of only 2–4 eggs. This may partly explain why barn owls rather than other *Tyto* species successfully colonized the globe.

**Variation in clutch size**    In contrast to egg size, clutch size varies greatly within and between barn owl populations. This is not always the case in birds, with some species laying a fixed number of eggs (e.g. the Adelie penguin typically lays one egg per clutch) and others, such as the kestrel, showing only slight variation in clutch size despite living in similar habitats to the barn owl. For example, in one German population of barn owls, clutches contained on average just three eggs in one year but as many as seven eggs in another.

Although the maximum recorded number of eggs in a single clutch is eighteen in Germany and the Czech Republic, barn owls could potentially lay more eggs. Females produce approximately 25 ovarian follicles, only some of which are fertilized. The supernumerary follicles are resorbed, suggesting that the decision to lay extra eggs can be made relatively late – for instance if some of the laid eggs are accidentally lost. By removing eggs one after the other, French researchers induced a female to lay eighteen eggs in a single clutch. This means that clutch size can be influenced by environmental factors experienced not only before egg laying but also throughout the laying period.

**Determinants of clutch size**    In the barn owl, clutch size often varies seasonally, and clutches are larger in years with early laying. In France and Switzerland, the mean size of the first annual clutch increases from February to June, while the size of the second clutch decreases only slightly from May to August. This is concordant with the observation that second clutches are on

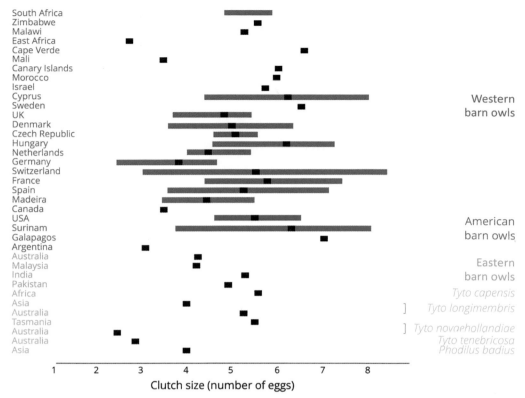

Mean clutch size is indicated by black boxes and the range in mean annual clutch sizes (where known) by horizontal coloured bars in the barn owl species complex (western, American and eastern barn owls), grass owls (*Tyto capensis* and *T. longimembris*), Australian masked owl (*T. novaehollandiae*), sooty owl (*T. tenebricosa*) and bay owl (*Phodilus badius*).

average larger than first clutches in most studied populations (Czech Republic, Denmark, France, Germany, Sweden, Switzerland, Malaysia and South Africa; see figure on page 160), including in a captive population in Germany to which food was provided *ad libitum*.

The only two exceptions were in populations in Spain and Utah (USA), where second clutches were smaller than first clutches, perhaps because in these two localities food supply decreases through the breeding season. In the Spanish Mediterranean region, the population density of wood mice decreases from early summer to winter as the vegetation dries out, depleting food resources for small mammals. In contrast, in central Europe, where second clutches are larger than first clutches, common voles reproduce in early spring, which boosts their numbers in summer followed by a decrease in autumn. This means that the size difference between first and second annual clutches is due to a seasonal effect rather than differing levels of investment between the two clutches. In other words, in a given population, two barn owls laying their first and second clutches on a similar date can be expected to lay a similar number of eggs.

The central role of food supply in explaining variation in clutch size in the barn owl has been demonstrated in Europe and Africa. Owls lay more eggs in peak than in low vole years, and in Scotland and Switzerland, heavier females lay more eggs than lighter females. The role of climate in determining clutch size has also been reported, but it remains unclear whether climatic conditions influence barn owl reproduction directly or indirectly through an effect on food supply. Current knowledge suggests that in Switzerland, France, the United Kingdom and the USA, clutch size is smaller after harsh winters than after mild winters, and in both Switzerland and the United Kingdom, clutches are larger during years when the laying period coincides with higher precipitation

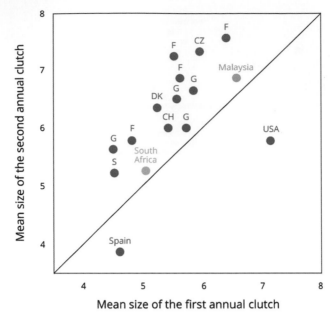

Mean number of eggs in the first and second annual clutches in barn owl populations in Europe (*orange circles*), Malaysia (*light blue circle*), USA (*dark blue circle*) and South Africa (*grey circle*). A circle positioned above the diagonal line indicates that second annual clutches are, on average, larger than first annual clutches, while the situation is reversed in Spain and the USA, which are located below this diagonal. CH, Switzerland; CZ, Czech Republic; DK, Denmark; F, France; G, Germany; S, Sweden.

levels, particularly when it rains more intensely three weeks before egg laying. Rain may boost vegetation growth and, in turn, small mammal populations, or higher amounts of rainfall may increase rodent activity and make them more vulnerable to predation.

The environment thus has clear effects on clutch size, but genetic factors still play a role, albeit a weaker one – with 20% of the variation in clutch size related to female identity. This contrasts to egg size, for which 80% of the variation is explained by female identity. Thus, although a given female lays eggs of similar size, the number of eggs can differ greatly between different reproductive attempts.

## FUTURE RESEARCH
• Assess the relative importance of genetic and ecological factors in determining clutch size.
• Determine whether females can predict and adjust clutch size in relation to the prey abundance that the offspring will be exposed to during the period of parental care and afterwards.
• Determine why clutch size in Europe has increased over the past fifty years.

## FURTHER READING
Béziers, P. and Roulin, A. 2016. Double brooding and offspring desertion in the barn owl (*Tyto alba*). *J. Avian Biol.* **47**: 235–244.

Chausson, A., Henry, I., Almasi, B. and Roulin, A. 2014. Barn owl (*Tyto alba*) breeding biology in relation to breeding season climate. *J. Ornithol.* **155**: 273–281.

Durant, J. M., Massemin, S. and Handrich, Y. 2004. More eggs the better: egg formation in captive barn owls (*Tyto alba*). *Auk* **121**: 103–109.

Moreno, S. and Kufner, M. B. 1988. Seasonal patterns in the wood mouse population in Mediterranean scrubland. *Acta Theriol.* **33**: 79–85.

Pinot, A., Gauffre, B. and Bretagnolle, V. 2014. The interplay between seasonality and density: consequences for female breeding decisions in a small cyclic herbivore. *BMC Ecol.* **14**: 17.

Roulin, A. 2002. Short- and long-term fitness correlates of rearing conditions in the barn owl. *Ardea* **90**: 259–267.

Toms, M. P. 1996. An eye to the future. *BTO News* **207**: 12–13.

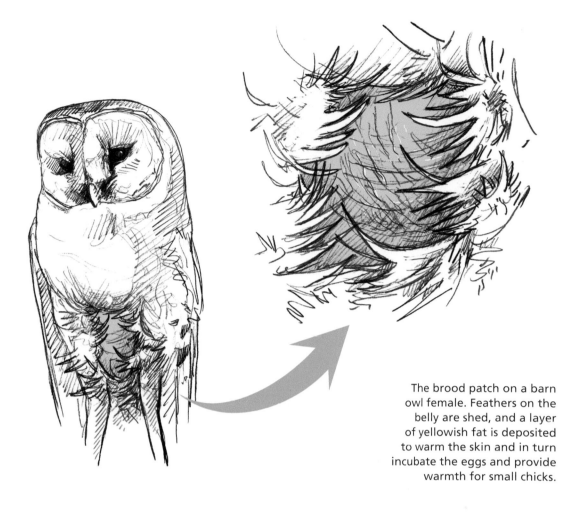

The brood patch on a barn owl female. Feathers on the belly are shed, and a layer of yellowish fat is deposited to warm the skin and in turn incubate the eggs and provide warmth for small chicks.

## 8.6   INCUBATION

# Broody hen

**In the barn owl, each sex plays a specific role in reproduction. The female incubates the eggs, warms the offspring and distributes the food collected by the male among the nestlings. Tytonidae eggs must be incubated for approximately 32 days before they hatch.**

**Incubation time**   For a bird's egg to hatch, it must be incubated for a period lasting anywhere from 13 days (in the canary) to 81 days (in the royal albatross). In the barn owl, this incubation lasts, on average, 32 days (range 27–36 days); in the grass owl, 29–31 days; and in the masked owl, an average of 33 days. The barn owl incubation period is approximately five days longer than that of similarly sized short-eared and long-eared owls. Is this because barn owl clutches are, on average, larger and thus require more energy and time for incubation? Or do barn owl mothers produce less warmth to incubate the eggs but do so for longer? Alternatively, do the mothers provide fewer resources for the eggs, which prevents barn owl embryos from growing as fast as the embryos of other owls? Whatever the reason, the length of incubation by the mother is not time-bound. If none of the eggs in her clutch hatches, she will continue to incubate them for as long as three or even six months.

**The brood patch and incubation behaviour** In Tytonidae, only the female has a brood patch, a zone on the belly where feathers are lost spontaneously before egg laying so that the eggs and hatchlings are in direct contact with the skin. Therefore, only the female incubates the eggs, although the male can on rare occasions assist by sitting on them. Without a brood patch, paternal help may not be very efficient.

During incubation the female usually remains inside the nest cavity, going out, on average, only 2.5 times per day. She can stay on the eggs most of the time because the male provides all the food she needs for self-maintenance and incubation. Barn owl females invest substantial energy in warming up the eggs, as demonstrated by the fact that 70% of the lipids found in the skin are concentrated in the brood patch. After the first egg has hatched, the mother leaves her nest more regularly. The quality of incubation may therefore decrease with the sequence in which eggs are laid. The eggs hatch asynchronously, and the last-laid eggs are possibly less assiduously incubated. This idea is supported by work carried out in Scotland, where 83% of the eggs that did not hatch were the last or penultimate eggs laid in a clutch. This effect might be even more pronounced in large clutches. The barn owl's brood patch is relatively small (about 12.8 square centimetres) and is not always large enough to incubate all eggs simultaneously. This could at least in part explain why, in Switzerland, the probability that one egg fails to hatch is greater in large than in small clutches.

Incubation is thus not a passive process, and females should monitor how they take care of each egg. This is probably what they do, as they change their incubating position by 180 degrees every minute to one hour. Female barn owls are very attentive to their eggs.

## FUTURE RESEARCH

- The duration of incubation and the incubation efficiency for each single egg in a clutch have rarely been accurately measured. It is therefore unknown whether duration is correlated with ambient and female body temperatures, clutch size, laying date and egg laying sequence.
- Incubation behaviour, including the assiduity with which females incubate the eggs and the time they invest in incubation, should be investigated.

## FURTHER READING

Boersma, P. D. 1982. Why some birds take so long to hatch. *Am. Nat.* **120**: 733–750.

Durant, J. M., Massemin, S., Thouzeau, C. and Handrich Y. 2000. Body reserves and nutritional needs during laying preparation in barn owls. *J. Comp. Physiol. B* **170**: 253–260.

Durant, J. M., Massemin, S. and Handrich, Y. 2004. More eggs the better: egg formation in captive barn owls (*Tyto alba*). *Auk* **121**: 103–109.

Durant, J. M., Gendner, J.-P. and Handrich, Y. 2010. Behavioural and body mass changes before egg laying in the barn owl: cues for clutch size determination? *J. Ornithol.* **151**: 11–17.

Epple, W. and Bühler, P. 1981. Eiwenden, Eirollen und Positionswechsel der brütenden Schleiereule *Tyto alba*. *Ökol. Vögel* **3**: 203–211.

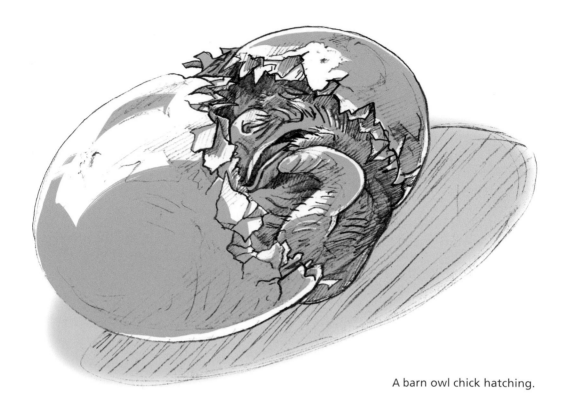

A barn owl chick hatching.

## 8.7  HATCHING

# Out of the box

**In Tytonidae, eggs hatch every 2–3 days, generating a pronounced size hierarchy, with the oldest nestling being up to one month older than its last-hatched sibling. First-born individuals have priority access to parental resources, and consequently, their last-hatched siblings are more likely to die when food supply is poor. Approximately 10–20% of eggs do not hatch, due to male infertility or early embryo death.**

**Hatching process**  Two days before hatching, the chick starts to vocalize inside the egg, probably to indicate an imminent need for food. Food is only brought back to the nest at night, but hatchlings require meals spread over 24 hours. To ensure access to a food stock, chicks therefore call to inform their parents about the timing of hatching, which takes between 12 and 36 hours. In response, fathers hunt more prey. The chick breaks the shell using an egg tooth located on top of the bill, and the mother often helps by removing large pieces of shell. Probably to prevent predators from finding eggshells that have fallen out of the nest, the father removes them, and the mother consumes them. It is, however, not rare to find eggshells in the nests, indicating that these parental behaviours are not essential and could depend on contextual cues, for example evidence of predator presence.

**Hatching success**  In clutches in which at least one egg hatched, 93% of the eggs hatched in Switzerland, 83% in Mali and 80% in Malaysia. No eggs hatched in only seven out of 933 nests.

Male infertility can be a major cause of hatching failure, as suggested in a Dutch study in which 55% of eggs showing no sign of embryonic development had sperm adhering to the shell membrane. Males were therefore able to transfer sperm that was, apparently, not able to fertilize the eggs. Male infertility may be more prevalent at older ages. A case study showed that until the age of nine years, a male reproduced nine times with six different females, producing a total of 55 eggs, 51 of which hatched (93%). However, from ten years of age onwards, the same male produced seven clutches, carefully incubated by two new partners, but only one of 44 eggs hatched (2%). A more thorough study in Switzerland showed that hatching success does indeed decline with male age, suggesting a decline in fertility with age.

Embryos can also die either because eggshells break accidentally or because of genetic problems. These deaths are apparently not due to variation in the quality of maternal incubation behaviour caused by environmental conditions. In Switzerland, the severity of the previous winter was not related to hatching success in the subsequent season. In addition, ambient temperatures, rainfall intensity and food supply during the breeding season did not predict the fate of eggs.

**Hatching asynchrony** Hatching asynchrony (i.e. eggs not hatching simultaneously) is the result of the female starting to incubate the eggs before she has completed the clutch, whereas hatching synchrony (i.e. eggs hatching simultaneously) is the outcome of the female not starting incubation until the clutch is complete. In so-called **precocial birds**, such as game birds and ducks, all eggs in a clutch hatch almost simultaneously. The already well-developed chicks can therefore walk out of the nest to venture to feeding places together. In **altricial birds** (most passerines, plus owls, raptors and many others) hatching synchrony is not so critical, because they stay in the nest for

A brood of six barn owl nestlings, showing pronounced size differences due to staggered hatching.

a time (from a few days in small passerines, up to several weeks in the case of the barn owl). In most altricial species, eggs in fact hatch more or less together, because incubation does not start until the clutch is complete. However, in owls and raptors asynchrony is the rule.

Some birds, such as the tawny owl and the kestrel, wait until a few eggs have been laid before starting to incubate them, but in the barn owl incubation starts as soon as the first egg is laid. This explains why the degree of hatching asynchrony is more pronounced in the barn owl than in many other owls and raptors. In this species, the age difference between siblings depends mainly on the time lapse between the laying of two successive eggs.

For barn owls studied in Argentina, the Czech Republic, France, the Galápagos Islands, Germany, India, Malaysia and the USA, grass owls in Australia, and masked owls in Tasmania, eggs are usually laid every 1–4 days, with the mean being 2–3 days. Nevertheless, a clutch can be laid in two cycles, sometimes up to 1–2 weeks apart, causing extreme age differences between siblings. In spite of the spread of laying dates, there is little variation in the duration of incubation. For instance, each of the seven eggs in a German clutch was incubated for a similar length of time, ranging from 31 days to 32 days and 7 hours.

Several hypotheses can explain the evolution of hatching asynchrony, and the **brood reduction hypothesis** is certainly the most popular. This states that during periods of food depletion or when parents have difficulties finding food, the last-hatched chicks, which are weaker than their older siblings, starve to death. The effect of hatching order on growth and ultimately survival is most pronounced during the most rapid phase of growth, between hatching and day 35. The death of the youngest chick or chicks is the best that could happen in such a situation, because parents will then have fewer mouths to feed. In Switzerland, in broods in which at least one nestling died, it was the later-hatched individuals of the brood in 95% of cases and the earlier-hatched individuals in only 5% of cases. If the eggs hatched simultaneously, siblings would all be the same age and would all have the same competitive ability, which would increase the risk of them all dying from starvation. Sacrificing part of the family to match brood size to food supply ensures that at least a few offspring can fledge successfully. This has been observed on numerous occasions in barn owls all around the world, indicating that hatching asynchrony has a clear adaptive function in this group of birds.

Although hatching asynchrony provides selective advantages to the parents when food is in short supply, it may potentially incur fitness costs under prime feeding conditions. When parents can find

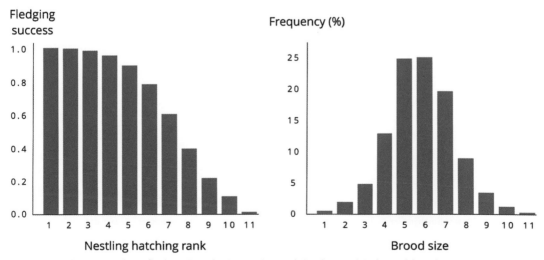

Probability that a nestling fledges in relation to its rank in the within-brood hatching sequence (*left panel*). For instance, 1 is for nestlings that are the first to hatch in a nest, and 3 is for the third-hatched chick. The right panel shows the frequency distribution of the broods of different sizes. The data are from Switzerland.

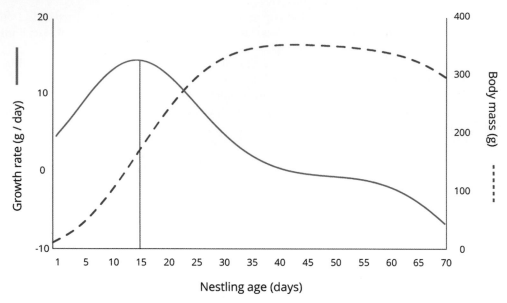

Nestling growth rate in terms of number of grams gained per day (unbroken orange line) and nestling body mass (dashed blue line) in Switzerland. The maximum growth rate occurs at 15 days of age, as indicated by the black vertical line.

enough food to avoid brood reduction, the physical superiority of senior nestlings allows them to monopolize a larger part of the parental resources at the expense of their younger siblings. This may increase the disparity in offspring body condition, ultimately impacting parental interest – because from the parents' point of view it may be better for all the offspring to have a similar start in life, rather than the first-born chick fledging in good condition and the last-born in comparatively poor condition. Costs and benefits of hatching asynchrony should be examined from the fitness perspective of parents, and of senior and junior nestlings.

Other hypotheses have been proposed to explain the adaptive function of hatching asynchrony. When siblings are of different ages, their peak in food demand, occurring at approximately fifteen days of age, is reached sequentially. Therefore, sibling competition is reduced, and parental workload is spread over time. Staggered hatching can also reduce the time parents spend at the nest, thereby minimizing predation risk. All these aspects have, however, not yet been evaluated in Tytonidae.

## FUTURE RESEARCH
- Data about the daily timing of hatching are required.
- The impact of the begging behaviour of the chicks before hatching on male feeding rate should be formally examined.
- The degree of hatching asynchrony should be recorded in multiple populations.

## FURTHER READING
Bühler, P. 1970. Schluphilfe-Verhalten bei der Schleiereule (*Tyto alba*). *Vogelwelt* **91**: 121–130.
Chausson, A., Henry, I., Ducret, B., Almasi, B. and Roulin, A. 2014. Tawny owl *Strix aluco* as a bioindicator of barn owl *Tyto alba* breeding and the effect of winter severity on barn owl reproduction. *Ibis* **156**: 433–441.
Roulin, A. 2002. Short- and long-term fitness correlates of rearing conditions in the barn owl. *Ardea* **90**: 259–267.
Stenning, M. J. 1996. Hatching asynchrony, brood reduction and other rapidly reproducing hypotheses. *Trends Ecol. Evol.* **11**: 243–246.

A brood of seven barn
owl nestlings.

## 8.8 BROOD SIZE

# A basket full of puppies

Barn owls, unlike masked owls and sooty owls, produce large broods. Food supply and weather conditions are key determinants of reproductive success. When the risk of predation is high or when competition for nest sites is intense, parents can desert the nest, particularly at the laying and incubation stages. If a reproductive attempt fails, many barn owl pairs produce a replacement clutch, particularly under prime feeding conditions and early in the breeding season.

**Variation in brood size**  Brood size varies greatly between populations and years, with the maximum observed number of fledglings in a brood being twelve in France and South Africa. Mean brood size is on average four nestlings in the western, American and eastern barn owls, and much smaller in masked owls and sooty owls, which have an average of 1–2 nestlings.

**Seasonal effect on brood size**  Losses of eggs and nestlings are frequent in the barn owl. For instance, in only 16% of nests in Utah did all eggs hatch, and all young fledge. In Europe, although clutch size usually increases through the season, brood size decreases. It seems that the number of eggs per clutch corresponds to feeding conditions at the time of laying and not to conditions prevailing a few months later, during the nestling-rearing period. Early in the season (around 1 March) in Switzerland, pairs lay 5.6 eggs, on average, because food is not yet available in large quantities. Food supply rapidly increases, explaining why as many as 5.2 of these 5.6 eggs result in a successful fledgling three months later. At that time and until the end of the summer, feeding conditions are better, and the last clutches, laid at the beginning of August, contain 7.2 eggs, on average. In the autumn, however, feeding conditions rapidly deteriorate, rain is more abundant, and ambient temperatures

A family of seven barn owl nestlings in Israel. © Amir Ezer

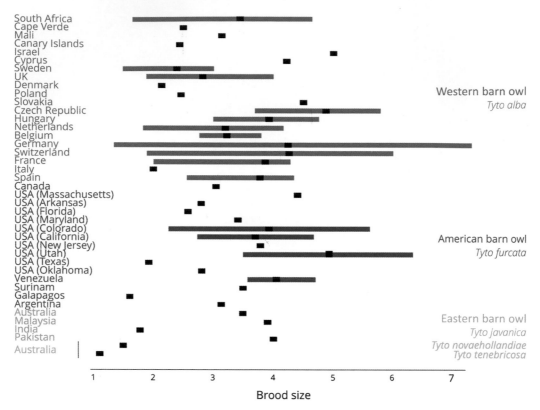

Mean brood sizes are indicated by black boxes, and ranges in mean brood sizes (where known) by coloured bars. Brood size seems to vary to a greater extent between years in Germany than in other countries; however, this conclusion may be biased because of the higher number of published results from Germany.

are lower, explaining why only 2.8 of the 7.2 eggs ultimately produce a fledgling. In arid regions such as Israel, reproductive success also decreases through the season because drought slows down reproduction in the owls' rodent prey.

## Impact of weather on brood size
As shown in Switzerland, although nestling mortality is more frequent in large than in small broods, the probability that at least one offspring reproduces the following year is still higher in large than small broods. Natural selection therefore favours optimistic parents who lay large clutches even if at that time they cannot predict how successful they will be in providing food to their offspring. For instance, parents hardly hunt during periods of heavy rain, which is particularly detrimental to young nestlings not yet strong enough to survive a few days without food. Precipitation is therefore a key determinant of reproductive success. However, the effect of weather can vary geographically, particularly with respect to ambient temperatures. In Switzerland, nestlings have a greater chance of survival in warmer years than in cooler years. At warm ambient temperatures (30 °C), nestlings can easily maintain a body temperature of 40 °C, whereas at cooler ambient temperatures (15 °C), young nestlings have difficulties warming up, and their cloacal temperature, on average, is only 34.5 °C. In Israel, the problem for nestlings is not cool ambient temperatures but rather heatwaves, during which they can die from overheating. The life of a young bird is not easy, wherever in the world it has hatched.

## Breeding failure
One study in Mali found that only 57% of breeding pairs produced at least one fledgling, but between 71% and 95% of pairs are successful in North America, the Middle East

A starving barn owl nestling.

and Europe. More than two-thirds of breeding failures occur at the incubation stage, because parents are more likely to abandon eggs than nestlings, particularly before the clutch is complete. Failure at the egg stage results from competition with conspecifics and other animals to secure the limited nest cavities that are safe from predators, and from humans who visit nests during the daytime at egg laying. The presence of humans at egg laying is considered a risk, and parents may abandon their clutch to find a safer site where they produce a replacement clutch.

Nest abandonment is particularly frequent in years when environmental conditions are so poor that the offspring have little chance of surviving. For instance, in Scotland, only 5–10% of nestlings die in good vole years, in contrast to 45% of nestlings in poor vole years. Nest desertion by the female is also more prevalent in years preceding a poor year when few of the surviving nestlings will breed. This means that parents can predict, at least to some extent, whether the rearing conditions will be optimal and if food will be plentiful the following year. Many rodents, the staple prey of barn owls, show specific patterns of abundance within and across years. For instance, if the abundance of rodents decreases in one year, owls can expect that the feeding conditions the following year will be even worse. In such conditions, the value of the current offspring is rather low and parents are therefore less attentive to them.

The death of the male is the greatest threat to the survival of young nestlings. Without the male, the main food provider, the brood is likely to fail. If he dies when the nestlings are still young and require their mother to brood and to tear flesh from prey for them to eat, the mother cannot leave the offspring unattended to forage. If the male dies when the nestlings are older and hence able to thermoregulate by themselves and eat without maternal help, they may still starve if the female does not increase the feeding rate or if she has already deserted the family to start a second brood, often at a great distance from the first nest.

**Replacement clutches** After a failed reproductive attempt, owls can either wait until the following year to reproduce or lay a replacement clutch one or two weeks later. In double-brooded species such as the barn owl, replacing a first annual clutch is like producing a second annual brood. Therefore, the frequency of re-nesting after a breeding failure is similar to the frequency of second broods, which is 14–56% in France. These numbers, however, are an underestimate. Replacement clutches are often produced with a different mate at a different site from where the failed clutch was laid, which may be as far as 85 kilometres away, as observed in Germany. Many replacement clutches are thus located outside the study area, with the result that ornithologists fail to record them. Replacement clutches are not particularly successful. In the USA, only 50% of pairs that re-nest produce full clutches.

Two factors are particularly important in the decision to re-nest: time and food supply. In France and Germany, clutches are more frequently replaced in years when first clutches are laid early in the season; this explains why barn owls are fifteen times more likely to re-nest following a breeding failure at the first breeding attempt than at the second attempt of the year.

## FUTURE RESEARCH
- Brood size measurements should follow a standardized methodology, to permit comparisons between barn owl populations. Comparing brood sizes between populations is subject to errors for three reasons. First, nestling mortality differs greatly between years, implying that brood size estimated from the long-term monitoring of populations better reflects the general trend than does that estimated from short-term monitoring. Second, brood size recorded soon after hatching overestimates the number of nestlings that fledge. Because the age at which nestlings are counted is rarely mentioned, comparing brood size between studies is not precise. Finally, to calculate mean brood size, researchers rarely distinguish pairs that failed to produce any fledglings from pairs that fledged at least one offspring.

## FURTHER READING
Altwegg, R., Schaub, M. and Roulin, A. 2007. Age-specific fitness components and their temporal variation in the barn owl. *Am. Nat.* **169**: 47–61.

Baudvin, H. 1986. La reproduction de la chouette effraie (*Tyto alba*). *Le Jean le Blanc* **25**: 1–125.

Chausson, A., Henry, I., Almasi, B. and Roulin, A. 2014. Barn owl (*Tyto alba*) breeding biology in relation to breeding season climate. *J. Ornithol.* **155**: 273–281.

Chausson, A., Henry, I., Ducret, B., Almasi, B. and Roulin, A. 2014. Tawny owl *Strix aluco* as a bioindicator of barn owl *Tyto alba* breeding and the effect of winter severity on barn owl reproduction. *Ibis* **156**: 433–441.

Dreiss, A. N., Séchaud, R., Béziers, P., Villain, N., Genoud, M., Almasi, B., Jenni, L. and Roulin, A. 2016. Social huddling and physiological thermoregulation are related to melanism in the nocturnal barn owl. *Oecologia* **180**: 371–381.

Kniprath, E. and Stier, S. 2008. Schleiereule *Tyto alba*: Mehrfachbruten in Südniedersachsen. *Eulen-Rundblick* **58**: 41–54.

Marti, C. D. 1994. Barn owl reproduction: patterns and variation near the limit of the species' distribution. *Condor* **96**: 468–484.

Roulin, A. 2002. Short- and long-term fitness correlates of rearing conditions in the barn owl. *Ardea* **90**: 259–267.

Rytkönen, S. 2002. Nest defence in great tit *Parus major*: support for parental investment theory. *Behav. Ecol. Sociobiol.* **52**: 379–384.

A barn owl nestling
close to taking its
first flight.

## 8.9   NESTLING GROWTH

# From chick to fledgling

Compared to similarly sized species, nestling barn owls grow slowly. Curiously, they accumulate an overshoot of body mass that they naturally lose before fledging. Once they start to roost outside their nest, they still return to eat and sleep. Juveniles definitively abandon the nest at 2–3 months of age, and become independent from their parents at 3–4 months of age.

**Obesity and weight loss**   At 40–45 days of age, nestling barn owls reach a body mass greater than that of adults. In Switzerland, the maximum male body mass is 470 grams in nestlings and only 412 grams in adults. Maximum nestling body mass varies between populations: 447 grams in India, 455 grams in the United Kingdom, 475 grams in France, 550 grams in Venezuela, 632 grams in Malaysia and 650 grams in the USA. After having reached such mass, nestlings naturally lose body mass – for instance, a barn owl chick in France loses 73 grams, on average, to reach 315 grams at fledging.

Why store extra body mass to lose it two weeks later? Initially, ornithologists assumed that the excess was reserve fat in case of food shortage, but it was later found to be due to water accumulation. But then the discovery that weight loss before fledging results from a reduction in food consumption, rather than from water loss, makes understanding the adaptive function of the surplus body mass difficult. Is it a way to fill the body with water like a tank, or to reduce food consumption just before fledging, or is it a reserve to boost the maturation of feathers, organs, tissues or physiological functions, such as the immune system? What we know for sure is that at fledging, all excess body mass has been used, and although better-fed nestlings gain more surplus body mass than do poorly fed nestlings, they all fledge with a relatively low body mass. The best situation for a nestling is therefore to accumulate extra body mass and to be slim at fledging.

**Determinants of nestling body mass** The abundance of the staple prey of the barn owl is the most important determinant of nestling body mass. Food supply varies between years, and so does nestling body mass. Pairs breeding early in the season also have access to more food than those breeding later, as observed in Mali and Switzerland, where, on average, nestlings weigh 371 grams on 15 April and only 343 grams on 15 September. In Switzerland, parents producing large broods are better able to feed each offspring than parents producing small broods. This implies that in high-quality habitats, i.e. those with less intensively cultivated open areas and without livestock, parents have access to enough resources to produce many high-quality offspring. This is not the case in Malaysia, where nestlings in large broods fare worse than those in small broods.

Although food abundance usually does not vary from one day to another, parents can sometimes have difficulties catching their prey. In Switzerland, the amount of rain that fell the previous night, but not two nights before, negatively affects nestling body mass. This suggests that parents can quickly offset the negative impact of a rainy night. For example, they may increase feeding rate after a short period of poor hunting conditions.

**Leaving the nest** Barn owls start to leave their nest for the first time at the age of approximately 55 days in Europe, 59–65 days in Malaysia, 61 days in Venezuela and 62–67 days in the USA and Argentina. Despite being much larger than barn owls, Australian masked owls take their first flight at a similar age, 60–65 days.

Some individuals are more eager than others to explore the outside world. In Switzerland, radio-tracked juveniles roosted outside the nest for the first time between the ages of 54 and 105 days (74 days on average). These ventures can be considered excursions, because the juveniles explore up to 4 kilometres from their nests, and two-thirds of the wanderers that roost one or several days outside their nest regularly come back to roost again. They definitively abandon their nest at the age of 60–113 days (79 days on average), and abandon the area surrounding the nest at 90 days, on average.

**Independence from parents** Offspring request more attention than parents are willing to give. For example, parents can abandon their offspring before the youngsters have learned to hunt. To evict their offspring, barn owl parents hiss at them and attack them, particularly if the parents initiate a second breeding attempt. This can be necessary because the agitated fledglings are noisy, disturb their mother during egg laying and incubation, and can usurp prey items destined for their mother. If parents do not plan a second clutch, they may tolerate and feed their offspring for longer, up to two months after fledging in the barn owl and up to 3–5 months in the larger sooty owls. If the mother plans a second annual breeding attempt, she stops feeding the offspring at the first nest three weeks before laying the second clutch.

## FUTURE RESEARCH

- The amount of body-mass overshoot should be quantified in different populations distributed throughout various ecosystems to examine which environmental factors may have promoted the evolution of short-term obesity.
- Parents and offspring should be tracked using GPS to determine the exact age when parental care stops.
- The fact that females sometimes desert their young offspring to produce a second annual brood with another partner therefore raises the possibility that females stop caring for their offspring before males do.
- Feeding rate should be recorded before, during and after rainy nights to examine how parents can offset the negative impact of a reduction in feeding rate following intense precipitation.

## FURTHER READING

Almasi, B. and Roulin, A. 2015. Signalling value of maternal and paternal melanism in the barn owl: implication for the resolution of the lek paradox. *Biol. J. Linn. Soc.* **115**: 376–390.

Almasi, B., Béziers, P., Roulin, A. and Jenni, L. 2015. Agricultural land-use and human presence around breeding sites increase stress-hormone levels and decrease body mass in barn owl nestlings. *Oecologia* **179**: 89–101.

Bendel, P. R. and Therres, G. D. 1993. Differential mortality of barn owls during fledging from marsh and off-shore nest sites. *J. Field Ornithol.* **64**: 326–330.

Bühler, P. 1980. Die Lautäusserungen der Schleireule (*Tyto alba*). *J. Ornithol.* **121**: 36–70.

Chausson, A., Henry, I., Almasi, B. and Roulin, A. 2014. Barn owl (*Tyto alba*) breeding biology in relation to breeding season climate. *J. Ornithol.* **155**: 273–281.

Csermely, D. and Sponza, S. 1995. Role of experience and maturation in barn owl predatory behavior. *Boll. Zool.* **62**: 153–157.

Durant, J. M. and Handrich, Y. 1998. Growth and food requirement flexibility in captive chicks of the European barn owl (*Tyto alba*). *J. Zool.* **245**: 137–145.

Durant, J. M., Landys, M. M. and Handrich, Y. 2008. Composition of the body mass overshoot in European barn owl nestlings (*Tyto alba*): insurance against scarcity of energy or water? *J. Comp. Physiol. B* **178**: 563–571.

Epple, W. 1985. Ethologische Anpassungen im Fortpflanzungssystem der Schleireule (*Tyto alba* Scop., 1769). *Ökol. Vögel* **7**: 1–95.

A barn owl mother incubating the eggs of her second annual breeding attempt. At the back are the nestlings from the first clutch.

## 8.10 SECOND AND THIRD ANNUAL BROODS

# Better twice than only once

**Although producing a single brood takes approximately 4–6 months, barn owls can breed two or three times in a year. This sensational reproductive rate is unique in owls and raptors. On average, 20% of barn owls in Europe produce two annual broods. The second clutch is typically laid even before the offspring of the first nest have fledged. High-quality individuals can produce a second annual brood only if they have time and sufficient food at their disposal and do not already have a large first brood to take care of.**

Raising more than one brood in quick succession is usually restricted to birds that have enough time within the season to produce several broods, i.e. small bird species with short periods of parental care. Double brooding is therefore rare in raptors and owls – but the barn owl is an exception. Given the very long period of parental care in this medium-sized species, the barn owl is a real oddity in the avian world.

**Reproductive potential**  In captivity, barn owls can produce up to five or six clutches in a row, the Australian and Tasmanian masked owls up to three broods, and the sooty owl up to two broods. In nature, raising three broods successfully one after the other in the same year is rare, and barn owls that lay three successive clutches often do so because the first and/or the second clutch failed to produce any fledglings. Cases of two successful annual broods have however been observed around the globe. One study reported 78% of pairs successfully producing two broods in one year in Malaysia, 56% in California, 32% in Venezuela and 11% in Utah. In Germany and Switzerland, up to 65% of the breeding birds may produce a second brood in some years, but fewer than a quarter of owls do so at least once in their lifetime.

Barn owls have nearly the same laying potential as that of chickens! In two or three annual clutches, females laid up to 32 eggs in Malaysia, 28 eggs in Germany, 24 eggs in France and 23 eggs in Switzerland. Although second annual breeding attempts are usually less successful than first ones, double-brooded pairs have in total more fledglings than single-brooded pairs. In Switzerland, for example, double-brooded pairs produced an average of 7.3 fledglings (sum of the fledglings produced from the first and second broods), compared with 3.9 for single-brooded pairs, and in France and the Czech Republic they reared a maximum of 17 fledglings.

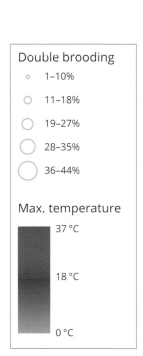

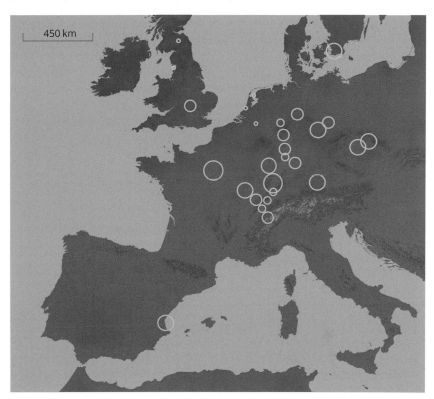

Mean frequency of double brooding in the European barn owl. For example, a percentage of 36–44% indicates that, on average, 36–44% of the breeding pairs produce a second annual brood.

Double brooding is therefore a strategy to increase the total number of offspring raised in a single year rather than simply to compensate for poor reproductive success in the first nesting attempt. In fact, the number of offspring raised during the first nesting attempt does not influence whether females produce a second brood. Additionally, being involved in a second breeding attempt does not jeopardize the success of the still-ongoing first nest, but double-brooded fathers must share prey items between two families, causing a slight reduction in body mass in the offspring in the first nest.

## Determinants of double brooding    Four factors explain variation in the propensity to be multi-brooded:

- **Food supply.** Owls more often produce a second annual brood in years when food is abundant. In a French study, in years when many pairs initiated two broods, their first annual broods contained 4.3 chicks, on average, and in years without second broods, the first annual broods contained only 3.0 chicks, indicating lower food abundance. Similarly, only the best sites with abundant food supplies could support two annual broods. This explains why sites where two emerge in the same year are re-used the following year more often than are sites with only one annual brood.
- **Time.** Raising two broods in a row is time-consuming, and therefore only pairs that produce their first brood early in the season have a chance to lay a second clutch. In France, Switzerland

and the United Kingdom, double-brooded owls laid their first clutch one month earlier, on average, than did single-brooded owls.

- **Parental duties.** Because second clutches are laid before the offspring of the first brood are independent, the capacity of the male to take care of two broods simultaneously can limit the likelihood of initiating another breeding attempt. As shown in Switzerland, having a large family prevents males from producing a second brood, because the food that they collect must be shared between two families, which slows the development of the offspring in the first nest. While males are stuck with a large first brood, females can desert their not-yet-independent first offspring to produce a second brood with another male free of parental care. Polyandry is rewarding!

- **Female quality.** Higher-quality females are more likely to lay a second annual brood than are poor-quality females. Accordingly, in Germany, Switzerland and Utah, compared to single-brooded females, those who produce two annual broods have a higher survival rate and are reproductively superior, producing larger first annual clutches.

## Laying interval between two annual broods

In Europe, the first and second clutches are initiated at 100-day intervals, on average. Some pairs can shorten this interval to 45 days, while others can wait up to 185 days. However, time is of the essence, and offspring must fledge and become independent before the weather gets too harsh and food resources are depleted. The necessity of not producing a second brood too late in the season selects for behaviour that can speed up the initiation of a new breeding attempt:

- **Hasty flirt.** In a captive barn owl population in Germany, courtship behaviour started 29 days before laying the first clutch (range 1–41 days) but only 18 days before laying the second clutch (range 10–23 days), probably because the nesting site was already established.

- **Change site.** By moving to another site, pairs save two weeks, as observed in Switzerland and the United Kingdom where a large number of nest boxes have been put up, which facilitates the change of nest site. If the mother stays inside the same small nest, she often has to wait until the offspring fledge before resuming egg laying. For pairs who do not break up to initiate a second annual breeding attempt, the new site can be close to the first nest (a few metres) or relatively far from it (up to 3 kilometres).

- **Change mate.** If a second nest site is not available close to the first nest, or if the partner is not yet ready to initiate a second brood, the last possibility for a female to initiate a second brood is to desert her first family and find a new partner free of parental care duties. By doing so, a female can lay the second clutch ten days earlier than if she had to wait until her first mate's provisioning responsibilities for the first nest were reduced. Approximately half of females choose this option in Switzerland.

- **Trade-off with the first brood.** Fledging many offspring takes time, and for this reason the laying interval between two annual broods is shorter if the first nest contains few rather than many offspring.

## FUTURE RESEARCH

- Climate warming causes animals to reproduce earlier in the season. This could increase the duration of the breeding season and thus provide more opportunities for barn owls to produce multiple broods in the same year. This aspect should be investigated, particularly in areas where ornithologists already possess long-term estimates of how many pairs produce two annual broods. This would offer an opportunity to investigate changes in the rate of double brooding with climate warming.

## FURTHER READING

Baudvin, H. 1986. La reproduction de la chouette effraie (*Tyto alba*). *Le Jean le Blanc* **25**: 1–125.

Béziers, P. and Roulin, A. 2016. Double brooding and offspring desertion in the barn owl (*Tyto alba*). *J. Avian Biol.* **47**: 235–244.

Kniprath, E. and Stier, S. 2008. Schleiereule *Tyto alba*: Mehrfachbruten in Südniedersachsen. *Eulen-Rundblick* **58**: 41–54.

Marti, C. D. 1997. Lifetime reproductive success in barn owls near the limit of the species' range. *Auk* **114**: 581–592.

A barn owl mother deserting her nest to produce a second annual clutch with another partner.

## 8.11 OFFSPRING DESERTION

# Dating younger males

**Deserting the family before the offspring are independent and letting the male complete parental duties allows female barn owls to accelerate the production of a second annual brood. To initiate a new breeding attempt, they attract a new, often inexperienced mate, sometimes at a great distance from the first nest.**

In mammals, internal gestation followed by lactation means that females are, literally, left holding the baby. It is therefore easier for males to desert when their care is not needed. In the barn owl, as in a few polyandrous birds, such as some waders and the rock sparrow, the opposite situation prevails. Males have no other choice than to care for their offspring, while the presence of the mother is not always mandatory at the end of the rearing period. Consequently, when the male partner is unable to support a second brood after the first one, the female may desert her family in search of new mating opportunities. The decision to stay or leave depends on the probability of finding a new mate and on whether the presence of the female is required to successfully raise the chicks.

**Why do females desert their offspring?**   In Germany, two-thirds of double-brooded female barn owls produced their two annual broods with the same partner, and the other third deserted the first nest to re-mate. In Switzerland, the situation is rather similar, with half of females and a quarter of males producing their two annual broods with different partners. Males can therefore change partners, but in contrast to females, they do not desert the first nest halfway through the rearing period. Nest desertion therefore provides females more opportunities to produce a second annual brood – with, on average, 13% of them being double-brooded, compared to 8% of males.

Deserting and non-deserting females are of similar intrinsic quality, and it is therefore mainly extrinsic factors that determine whether an individual abandons her family:

- **Time constraints.** A female can search for a male free of parental duties to quickly initiate a second breeding attempt rather than wait for her mate to have less work with the first brood. Because environmental conditions deteriorate at the end of the season, late breeders are more likely to desert than early breeders, to speed up the production of a second brood before it is too late.
- **Large family.** Males, who oversee the feeding of the offspring, are even more constrained when they produce many offspring in the first breeding attempt. The impatient female may desert the family to save time, and find a new male with no parental responsibilities. Among double-brooded females in Switzerland, those who deserted had, on average, 4.5 offspring in the first nest, while non-deserting females had 3.8 offspring.
- **Prevent males from sharing their investment between two families.** From the female's perspective, it might be better if her male is not involved in two nests, to avoid sharing food between two families. By deserting, a female may thus ensure that her first male devotes all his energy to her first brood. This seems to be an efficient strategy, because fewer than 23% of males whose partner deserted the family had the opportunity to raise a second family with another mate.
- **Annoying offspring.** The offspring from the first nest can disturb parents while they are copulating and chase the father to obtain a prey item intended for their mother. This might induce the mother to desert the nest to quietly start a second brood elsewhere.
- **Temptation.** Males who could not attract a female to breed early in the season may be desperately looking for females. If floater males are numerous, a female may have more opportunities to desert her first family.

## Who is the lucky male chosen by a deserting mother?
Finding a new male takes time, and a female may travel long distances, going back and forth from her first nest. In Switzerland, one female flew 60 kilometres and two days later returned to her nest empty-handed. Females usually attract a new male who is free of parental responsibility, as in 77% of cases the female's second brood was the male's first. A further 5% of cases represented a replacement clutch after first-brood failure. Only in 18% of cases was it the male's second annual brood. This is good news for the deserting female, because her new partner can devote all his energy to the new brood. The trade-off is that the female must abandon an experienced male (83% of the males had already bred in previous years) to seduce an inexperienced yearling individual (50% of them had never bred before).

## Fitness consequences of nest desertion
Female desertion implies that her male must alone fulfil parental duties for the first nest. Desertion often occurs when the offspring are still very young, with the last-hatched nestlings as young as 22 days old, an age at which the nestling is still highly dependent on its mother for brooding and feeding. Surprisingly, these duties are, in part, taken up by the oldest nestlings, who huddle with their youngest siblings and feed them. Did maternal selfishness promote the evolution of cooperative behaviour among young siblings?

Deserting the chicks is mainly a matter of saving time. The number of days between laying the first and second clutches was 83 days in deserting females, 91 days in non-deserting females that changed nest box to produce the second annual brood, and 103 days in non-deserting females that did not change box. Although deserting females produced more eggs than did non-deserting females in the second nest, brood size and offspring body mass were similar in deserting and non-deserting females. Thus, females desert their brood not because the first mate is of poor quality but often simply to accelerate the production of the second annual brood. Supporting this claim is that half of the females who deserted their brood to embark on a second breeding attempt returned to their previous mate to raise their first annual brood the following year. There is no bitterness!

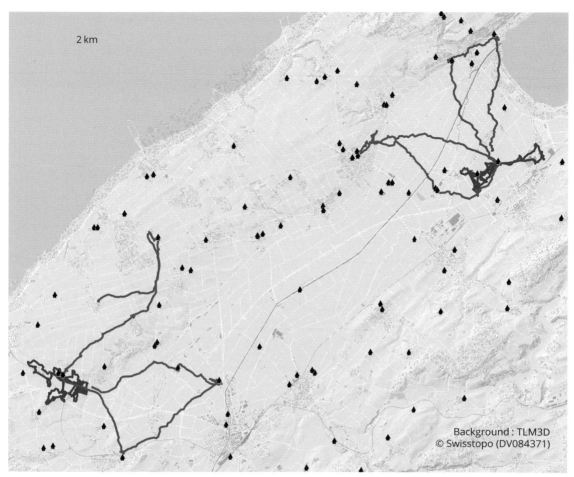

Male (*blue lines*) and female (*orange lines*) movements at two nests containing one-month-old offspring in Switzerland. The two males forage relatively close to their nests, while their female partners visit other nest boxes (*black triangles*) in search of another site and potential partner to initiate a second annual brood. Data obtained with GPS fastened on the backs of breeding barn owls.

**Desertion implies changing site** The nest site and surrounding territory belongs to the male (though once a female has decided to lay eggs, she tenaciously occupies the nest site while the male occupies the surrounding territory). Therefore, when first and second annual broods are found at a given site, they always belong to the same male but not necessarily to the same female. In a quarter of cases, a double-brooded male has changed mate to produce the second annual brood. Is this because his previous female deserted the first nest and was therefore not available for the second brood?

Regardless of whether a male changes his mate, in more than half of cases he produces his two annual broods at different sites. The new site is still in his territory, not very far from the first nest (from a few metres up to 3.3 kilometres) where his not-yet-independent offspring still require food provisioning. In contrast, females travel longer distances to breed with another male. They might have to nest relatively far from the first male (on average 4.6 kilometres, with a maximum of 41 kilometres). Nevertheless, sometimes a male allows his unfaithful mate to initiate a second brood with another male, even less than 40 metres from his nest. Is this tolerance traded for continued help from the female at the first nest, or does the first male agree to the new arrangement if his deserting female visits him to copulate?

**Copulating with the deserting female** If a male cannot prevent his mate from abandoning the nestlings, he may still be able to sire some of her new offspring. Pairs often copulate while raising the chicks of the first brood, which results in 8% of second broods of deserting females containing young sired by the first male. These females bred relatively close to their first nest (at up to 1500 metres), suggesting that they continued to visit their first male, as observed once in Switzerland, or that the first male joined his previous female close to her second nest. For males who cannot support a second annual brood, female desertion may be an opportunity to sire a few extra offspring at her second nest. Females may also benefit if the first male is of higher genetic quality than the new mate.

## FUTURE RESEARCH

- Do females desert their first nest to produce a second annual brood with a new mate (serial polyandry) more often than males successfully attract two females simultaneously (polygyny)?
- If a female has initiated a second annual brood with a new male, does she nevertheless continue to take care of her offspring at the first nest?

## FURTHER READING

Béziers, P. and Roulin, A. 2016. Double brooding and offspring desertion in the barn owl (*Tyto alba*). *J. Avian Biol.* **47**: 235–244.

Dreiss, A. N. and Roulin, A. 2014. Divorce in the barn owl: securing a compatible or better mate entails the cost of re-pairing with a less ornamented female mate. *J. Evol. Biol.* **27**: 1114–1124.

Henry, I., Antoniazza, S., Dubey, S., Simon, C., Waldvogel, C., Burri, R. and Roulin, A. 2013. Multiple paternity in polyandrous barn owls (*Tyto alba*). *PLoS One* **8**: e80112.

Kniprath, E. and Stier, S. 2008. Schleiereule *Tyto alba*: Mehrfachbruten in Südniedersachsen. *Eulen-Rundblick* **58**: 41–54.

Kniprath, E. and Stier, S. 2011. Umstände des Partnerwechsels für eine Zweitbrut der Schleiereule *Tyto alba*. *Vogelwarte* **49**: 75–77.

Kokko, H. and Jennions, M. D. 2008. Parental investment, sexual selection and sex ratios. *J. Evol. Biol.* **21**: 919–948.

Roulin, A. 2002. Offspring desertion by double-brooded female barn owls *Tyto alba*. *Auk* **119**: 515–519.

# 9 Parental care

9.1  Parental foraging
9.2  Parental behaviour at the nest
9.3  Adult body mass
9.4  Food stores
9.5  Adoption

Hunting a vole.

## 9.1  PARENTAL FORAGING

# The hunt is on

**As in other owls and raptors, the male barn owl is mainly responsible for bringing food back to the nest. During the first half of the rearing period, the mother distributes prey items among the nestlings. Once the offspring become older, the mother must decide whether she assists the father. Usually, she delivers about a quarter of the prey items, but in some families she does not participate in hunting.**

Why does one parent invest more in foraging for the nestlings? Do the parents negotiate investment? Can they compensate for their mate's poor performance? Are parents sensitive to offspring begging, or do they invest a fixed budget in parental care, regardless of offspring demand? These are key issues for understanding how animals allocate time and energy for parental care.

**Parental feeding rates**  Foraging success can sometimes be extremely high in the barn owl, as illustrated by a male in the United Kingdom who brought back eleven prey items in under an hour. The feeding rate, however, is usually more modest. In Switzerland, in broods of on average five chicks, parents bring a mean of eleven prey items per night, with a maximum of 31 items. Variation in feeding rate depends on many factors, such as habitat quality, fluctuations in prey populations, timing of prey activity, brood size and parental quality. When a hunt is successful the parent returns to the same spot, as observed in France. Weather conditions have an important impact, as males rest for over 90% of rainy nights instead of hunting.

A full picture of parental workload would also require knowledge of prey size, yet this has rarely been considered. This parameter is important, however, as illustrated by a study in India showing that parents who catch small prey items deliver more of them per night than do parents who catch larger items.

## Parental roles in feeding

Males bring more prey items to their mate during egg laying than during incubation, probably because producing eggs is energetically more demanding than incubating them. Once the eggs start to hatch, the male increases the feeding rate. However, when the family becomes older and therefore hungrier, the male may only slightly augment the feeding rate, probably because he has reached the peak of his foraging abilities. In Europe, while the mother may also forage for the brood, the father still brings most of the prey items to the nest: three-quarters of all items, on average.

If the father cannot meet increasing food demands, the mother resumes hunting, usually fifteen days (range 14–17 days) after the hatching of the first egg. She must lose weight to be more agile, a process that is initiated six days before she returns to hunting. In France, once the first-hatched nestling is on average 27 days old and the youngest 20 days old, the mother starts to spend entire nights outside the nest cavity except to deliver food. If the male can supply all the food required by the family, however, she continues to take care of the younger chicks in the nest, instead of hunting. In Switzerland, in 103 nests that were monitored over 259 nights, females did not bring any prey items on 31% of the nights, leaving their mate to do the job. At one nest in France, the female never hunted, probably because the male was able to provide enough food for all the chicks.

## Timing of feeding activities

Foraging success relies on agility and on the reduction of flying costs. To deliver many prey items to their hungry families, barn owls strive to minimize body weight, which they accomplish by regulating when and how much they eat. This is one of the reasons why, at the beginning of the night, the father brings the largest prey items to his offspring and has only a small meal himself before eating more food later in the night. This behaviour has been observed during the entire rearing period in Europe and North America, while in Venezuela barn owl parents have two peaks in hunting activity, delivering prey items mainly at the start and the end of the night. Such geographic differences might be due to the timing of activity of different prey species. When the young are less than two weeks old the mother, like the father, distributes food among the nestlings early in the night and feeds herself later on.

## Parental response to offspring begging

In all animals with parental care, offspring beg for food by means of conspicuous begging behaviour, such as loud vocalizations. Parents can adjust the feeding rate in response to this stimulus. In the barn owl, offspring solicitations may motivate the mother to assist the father in foraging for the brood. However, three observations point to the possibility that the main function of begging behaviour may not be to induce parents to increase foraging effort. First, parents provide most of the prey items necessary to fulfil the daily food requirement during the first few hours of the night. Thus, prey items accumulate more rapidly than offspring can consume them. This suggests that parents do not adjust feeding rate to the current food need of their offspring but somehow know how much food their offspring will consume over 24 hours. Second, parents raising a brood that had been experimentally enlarged or reduced by two chicks delivered food at similar rates. Thus, the degree of parental care may have been fixed before brood size manipulation, implying that parents may not increase hunting effort if offspring suddenly become hungrier. This interpretation is consistent with a third observation – that in nests to which ornithologists added extra food one day, the parents did not reduce the feeding rate the following night, resulting in the accumulation of larger food stores. Parents sometimes bring more food than necessary, which goes to waste.

Therefore, the main function of offspring begging may be to influence which individual receives a prey item once a parent returns with food. Given that the mother distributes the food stored in the

As in most raptors, the male hunts and the mother distributes food among the young.

nest, this might explain why the offspring direct their begging more intensely towards the mother than towards the father, who takes less time in choosing to which nestling he gives a recently captured prey item. This is plausible because, in Switzerland and the United Kingdom, nestlings that begged more intensely than their siblings were fed with priority by the parents irrespective of how close they were to the nest entrance.

## FUTURE RESEARCH

- If a female adjusts her feeding rate in relation to male feeding rate, and if the male is unable to deliver enough food to the brood, how fast is this adjustment? Is it a matter of hours, a day, or several days? Addressing this would allow us to investigate whether female compensation for food provisioning occurs only when the male cannot properly meet the brood's need over the entire rearing period, or if she also makes adjustments for temporary reductions in male feeding ability.
- Which ecological factors and individual characteristics determine when the mother starts to hunt for the offspring and stops brooding and distributing food that the male has brought?
- Because females start to participate in hunting when the last-hatched offspring are still young, is early maternal participation in hunting associated with increased mortality of the last-hatched chicks? In other words, is maternal participation in hunting associated with a higher survival rate of the oldest offspring at the expense of the survival rate of the last-hatched offspring?
- Parents forage for the brood in the first part of the night, before feeding themselves. Is this behaviour directly related to the quantity of prey that they need to catch? In other words, if the food requirement of the brood decreases, how do parents divide foraging time through the night?
- At nestling independence, does the mother stop caring for the offspring before the father?
- How the father and mother allocate food among the offspring should be recorded. The father takes less time to give a food item to one of his offspring; therefore, the mother and father may favour different offspring when allocating food. Does the father prioritize the older chicks, and does the mother give food to the hungriest nestlings? This aspect of parental care is still largely unexplored in the barn owl.

## FURTHER READING

Almasi, B., Roulin, A., Jenni-Eiermann, S. and Jenni, L. 2008. Parental investment and its sensitivity to corticosterone is linked to melanin-based coloration in barn owls. *Horm. Behav.* **54**: 217–223.

Durant, J. M. 2002. The influence of hatching order on the thermoregulatory behaviour of barn owl *Tyto alba* nestlings. *Avian Sci.* **2**: 167–173.

Durant, J. M., Gendner, J.-P. and Handrich, Y. 2004. Should I brood or should I hunt: a female barn owl's dilemma. *Can. J. Zool.* **82**: 1011–1016.

Durant, J. M., Hjermann, D. Ø. and Handrich, Y. 2013. Diel feeding strategy during breeding in male barn owls (*Tyto alba*). *J. Ornithol.* **154**: 863–869.

Epple, W. 1985. Ethologische Anpassungen im Fortpflanzungssystem der Schleiereule (*Tyto alba* Scop., 1769). *Ökol. Vögel* **7**: 1–95.

Langford, I. K. and Taylor, I. R. 1992. Rates of prey delivery to the nest and chick growth patterns of barn owls *Tyto alba*. In Galbraith, C. A., Taylor, I. R. and Percival, S. M. (eds) *The Ecology and Conservation of European Owls*. Peterborough: Joint Nature Conservation Committee, pp. 101–104.

Pande, S. and Dahanukar, N. 2012. Reversed sexual dimorphism and differential prey delivery in barn owls (*Tyto alba*). *J. Rapt. Res.* **46**: 184–189.

Roulin, A. and Bersier, L.-F. 2007. Nestling barn owls beg more intensely in the presence of their mother than in the presence of their father. *Anim. Behav.* **74**: 1099–1106.

Roulin, A., Ducrest, A.-L. and Dijkstra, C. 1999. Effect of brood size manipulations on parents and offspring in the barn owl *Tyto alba*. *Ardea* **87**: 91–100.

Roulin, A., Kölliker, M. and Richner, H. 2000. Barn owl (*Tyto alba*) siblings vocally negotiate resources. *Proc. R. Soc. Lond.* B **267**: 459–463.

Roulin, A., Da Silva, A. and Ruppli, C. A. 2012. Dominant nestlings displaying female-like melanin coloration behave altruistically in the barn owl. *Anim. Behav.* **84**: 1229–1236.

A male barn owl bringing a vole to his almost-fledged young in Israel. © Amir Ezer

## 9.2  PARENTAL BEHAVIOUR AT THE NEST

# The chef

The father is mainly responsible for hunting for the brood, while the mother prepares the meal by tearing the flesh into pieces for her young to eat. Although parents can transfer food to offspring anxiously blocking the nest entrance, they usually enter the nest cavity and first offer food to the hungriest individual, identified by intense begging solicitations. The two parents use different rules to feed the offspring.

Offspring request more attention than parents are willing to provide. Parents should not work to exhaustion, to avoid jeopardizing their own survival in the following years. Even if offspring have some interest in their parents producing siblings or half-siblings in the future (so-called 'kin selection'), their prime interest is to be in good enough condition to survive the first winter and reproduce the following year.

**Arriving at the nest** During the incubation period, the male visits the nest frequently, but not always with a prey item, as observed in a nest in France where a male brought a food item on only half of eight nocturnal visits. After hatching, the male usually brings a prey item for the young when he returns to the nest. When he does come without bringing prey – up to 20% of nest visits – is this to monitor offspring food need or to determine whether the female is present at the nest?

Carrying a heavy prey item over long distances is tiring, and solutions are needed to reduce the cost of flying. To maintain good balance during flight, the prey is held in the talons and, once the owl is perched near the nest, transferred to the bill and given to a nestling. Because the largest prey items can be half the weight of a breeding owl (up to 120 grams in Europe), the body part with the least amount of flesh can be

Parents tear apart pieces of flesh to feed their young offspring.

removed to reduce transportation load. In a Czech population, 33% of the items were decapitated before being delivered, whereas in Scotland prey items were smaller and always delivered intact.

Although decapitation facilitates the consumption of large items, the father probably does not deliver decapitated prey for this reason. If this were the case, he would also dismember large items and tear out flesh to help his offspring eat, something that only the mother does. Another possible reason for the delivery of decapitated prey might be that the parent consumes the head preferentially, as the brain tissues are rich in proteins, carbohydrates and lipids.

What is sometimes observed is the male picking up a prey item lying in the nest and giving it to soliciting offspring. In Switzerland, 14% of the prey items transferred by a parent to the offspring were already lying in the nest (the other 86% were freshly caught in the field). Although the nestlings could have taken these prey items from the nest floor themselves, this parental behaviour stimulates the offspring to eat. To further stimulate the chicks, parents are sometimes seen to pick up a prey item, leave the nest and return with it a few minutes later!

**Preparing the meal** When the offspring are still young, the father forages and transfers food items to the mother, usually inside the nest cavity. Even if some items remain uneaten there, the mother still begs for food from her mate, perhaps to motivate him to bring back more food for later consumption. The mother prepares the meal using the prey items brought by the father, removing the viscera and tearing up pieces of flesh. While feeding her young offspring, she produces chattering calls that induce begging. While it remains unknown whether she takes care of each hatchling in proportion to its begging behaviour, it is plausible – because she can feed a single offspring up to fifteen pieces of meat in a row before feeding another chick.

**Parental food allocation among the nestlings** When the offspring are 4–8 weeks of age, parents give prey items preferentially to the individual that vocalizes the most in 70% of cases. This occurs because at that age the nestlings that produce more and longer vocalizations are hungrier than their less vocal siblings. However, parents can still make errors, rectifying them subsequently, as illustrated by a mother who, just after giving a food item to a nestling, took it back and gave it to a sibling that was persistently begging. Such errors may be frequent in crowded broods because of the difficulty of accurately identifying which nestlings are making the most noise. Avoiding such errors requires taking the time to carefully assess begging calls. This contrasts with the quick food transfers characteristic of the mother. Thus, the offspring have more time to convince their father to give them a prey item, which perhaps explains why they beg more intensely for food from their mother (although for a shorter period) than from their father.

Although parents allocate food among the nestlings based on begging behaviour, they can also induce physical competition to favour the oldest and strongest offspring. The mother and father transfer prey items without entering the nest cavity in 25% and 12% of the visits, respectively, a behaviour that favours offspring which can monopolize the nest entrance without falling out of the nest. If the mother is less keen than the father to enter the cavity, this is apparently not because the offspring behave differently in front of their mother than their father. Indeed, the offspring cannot anticipate which parent will come next, as the sequence of maternal and paternal arrivals with food is random. Thus, the nestlings do not block the nest entrance more often at the arrival of the mother than at the arrival of their father. In summary, the mother and father use different rules to decide how they will deliver the food, and how to prioritize which offspring is fed.

## FUTURE RESEARCH

- In relation to the items that are caught, do parents bring the largest, medium-sized or smallest prey items to the nest and consume the others themselves? Small items may not be worth the flight back to the nest, whereas nestlings may have difficulties consuming large items without maternal help. Do parents select prey items for their young post-capture, deciding what to keep for themselves and what to bring back to the nest?
- Why do males sometimes return to the nest without any food item?
- Do parents adopt different foraging methods and strategies when their offspring are hungrier?
- Do parents decapitate large prey items depending on how far away from the nest they capture the prey?
- Compared to fathers, are mothers more likely to give a prey item to the offspring that begs the most? If nestlings beg less intensely in the presence of the father than the mother, is this because the father is less sensitive to offspring solicitation and feeds in priority the oldest offspring? More information on how parents allocate food among the nestlings is required.
- Just before entering the nest cavity, the father often produces a distinctive call, as observed in the barn owl in Switzerland and in the red owl in Madagascar. Why does it appear that the father, but not the mother, conspicuously informs the family that he is coming? Is it to prevent the offspring from begging as loudly as they do in the presence of the mother? This could be tested experimentally by playing back paternal calls just beside the nest entrance.

## FURTHER READING

Bühler, P. 1980. Die Lautäusserungen der Schleireule (*Tyto alba*). *J. Ornithol.* **121**: 36–70.

Roulin, A. and Bersier, L.-F. 2007. Nestling barn owls beg more intensely in the presence of their mother than in the presence of their father. *Anim. Behav.* **74**: 1099–1106.

Roulin, A., Da Silva, A. and Ruppli, C. A. 2012. Dominant nestlings displaying female-like melanin coloration behave altruistically in the barn owl. *Anim. Behav.* **84**: 1229–1236.

Thorstrom, R., Hart, J. and Watson, R. T. 1997. New record, ranging behavior, vocalization and food of the Madagascar red owl *Tyto soumagnei*. *Ibis* **139**: 477–481.

A barn owl resting before resuming parental care.

## 9.3   ADULT BODY MASS

# Heavy duty

Reproductive activities are so demanding that from the start of the incubation period to the end of chick rearing females lose an average of 114 grams (27% of their body weight), reaching a body mass below that of the non-breeding season. Males lose only a third as much, 30 grams (10% of their body weight). In both parents, minimum body mass is reached when the offspring are 40 days of age, some two weeks before fledging.

Breeding birds adjust body mass to the reproductive cycle. In France and Switzerland, female barn owls start to gain body mass on average 19 days before laying the first egg, and their body mass peaks (at about 430 grams) when the first egg is laid. It then progressively decreases, to reach about 316 grams when the oldest offspring of the brood is 42 days on average, before increasing continually until the end of the parental care period.

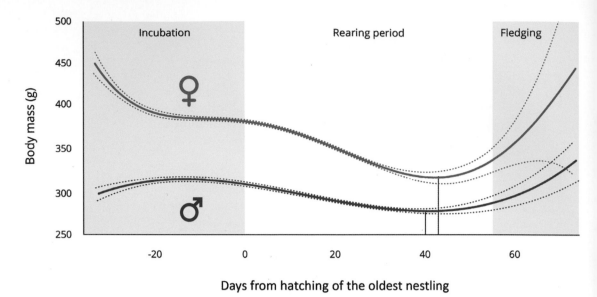

Body mass of 468 male and 620 female breeding barn owls in Switzerland in relation to the stage of reproduction. In each brood, the oldest nestling hatched on 'day 0' and took its first flight at about 55 days. Male and female breeding birds were weighed, on average, 2.2 times (maximum 11 times) and 4.4 times (maximum 38 times), respectively. Minimum body mass is indicated by a vertical black line (on average, 278 grams in males when the offspring were 40 days old and 316 grams in females when they were 42 days old).

By fasting from seven in the morning until midnight, females lose on average 12 grams. This daily loss is more pronounced during offspring rearing than during incubation, probably because, during daylight hours, more prey remains in the nest that can be eaten by the mother. When prey is available in the nest, females are heavier by exactly 12 grams compared to when food is not available. Not surprisingly, in years when prey is more abundant, female barn owls are heavier than in years when food supply is poor. This could partly explain why some individuals achieve a higher reproductive success than others: females with many offspring are heavier than those with fewer offspring.

Successfully raising a family also imposes requirements on males, who provide most of the prey items for their young. To increase agility and hence foraging success, males must lose body weight. The mean body mass of males is 308 grams during the incubation period and reaches 278 grams (a loss of 10% of body weight) when the first-hatched chick is 40 days old. Interestingly, males and females reach their lowest body weight at the same stage of reproduction.

## FUTURE RESEARCH

- The ecological factors that impact adult body mass should be identified. Because adults may spontaneously lose body mass independently of effort invested in reproduction, researchers should investigate whether body-mass loss is strategically regulated (i.e., adults lose body mass to enhance reproductive performance). Alternatively, rather than helping parents improve foraging efficiency, is body-mass loss the by-product of working very hard to raise a family?
- If body-mass loss during the rearing period is adaptive, it could be associated with reproductive success. As a test, male and female body mass should be recorded repeatedly throughout the entire reproductive season.

- Older females are heavier than younger ones (in Switzerland, an adult of a given age is, on average, 2.2 grams heavier than an individual one year younger). Is the difference in body mass between young and old breeding owls due to experience improving hunting efficiency, or to the ability to secure high-quality territories?
- Males and females reach their lowest body mass when their offspring are about 40 days old. Does this coincide with the brood's maximal food need? To answer this question, the total amount of food consumed by broods should be measured in relation to nestling age.

## FURTHER READING

Baudvin, H., Jouaire, S. 2001. Breeding biology of the barn owl (*Tyto alba*) in Burgundy (France): a 25 year study (1971–1995). *Buteo* **12**: 5–12.

Durant, J. M., Gendner, J.-P. and Handrich, Y. 2004. Should I brood or should I hunt: a female barn owl's dilemma. *Can. J. Zool.* **82**: 1011–1016.

Durant, J. M., Gendner, J.-P. and Handrich, Y. 2010. Behavioural and body mass changes before egg laying in the barn owl: cues for clutch size determination? *J. Ornithol.* **151**: 11–17.

Durant, J. M., Hjermann, D. Ø. and Handrich, Y. 2013. Diel feeding strategy during breeding in male barn owls (*Tyto alba*). *J. Ornithol.* **154**: 863–869.

Roulin, A. 2009. Covariation between eumelanic pigmentation and body mass only under specific conditions. *Naturwissenschaften* **96**: 375–382.

Uneaten prey items
lying in a nest.

## 9.4  FOOD STORES

# In the fridge

**Most raptors, owls and shrikes cache food for later use. Although storing food is beneficial in case foraging conditions deteriorate, barn owls are not so prudent. They store food in their nest but do so under prime foraging conditions and not necessarily when it might be most useful, such as before poor weather conditions. Prey items accumulate in the nests mainly because nestlings delay their meals, eating both night and day.**

Barn owls only occasionally cache food in their winter roosts, but they often accumulate prey remains in their nest when breeding. Why?

**Insurance against lack of food?**  Stocks of food are opportunistically built up when a good supply of small mammals suddenly becomes available. In France and Switzerland, barn owl parents store more prey in their nest in years when many pairs are breeding, i.e. years when food is plentiful. Do they store food in prime foraging conditions as an insurance against events such as inclement weather? No! In Switzerland, parental feeding rate does not decrease after food has been stored, and reproductive success is no higher if many prey remains are stored. Additionally, an owl in captivity, and therefore without any lack of food, killed 165 mice and hid many of them in nine distinct places inside the aviary.

A male barn owl accumulating prey in his nest cavity.

**Is it all waste then?** Unless food stores contain an exceptionally large number of prey remains, they are normally consumed in less than a day. However, some prey items lie untouched for days. These are usually the largest items that cannot be swallowed whole and represent more than one meal: to be consumed, pieces of flesh need to be torn off them, which is difficult for young barn owls. Indeed, nestlings can sometimes handle a large prey item for 45 minutes without successfully eating a single piece. Therefore, the smallest items are devoured first and the largest are either consumed later with the help of the mother or wasted. The observation that during the entire rearing period, fewer than two prey items rot and are never eaten indicates that parents do not kill more prey

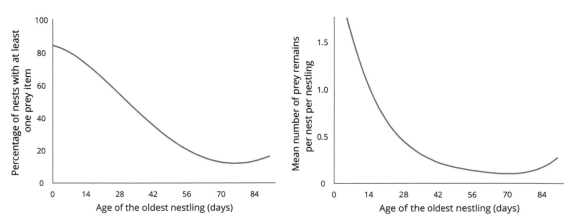

Frequency of uneaten prey items lying in barn owl nests in Switzerland in relation to the age of the oldest nestling. The graph on the left shows that at hatching, at least one prey item is found not yet eaten in more than 80% of nests. The percentage of nests decreases progressively to reach a minimum value of 15%. The graph on the right shows that at hatching, there are usually more than 1.5 uneaten prey items per nestling.

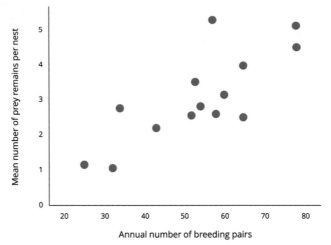

More prey remains are found not yet eaten in barn owl nests in years when more barn owls are breeding. Data were collected in Switzerland from 1990 to 2003.

than required each day. Therefore, if prey remains are found in barn owl nests during the daylight hours, it is not because this pile of food is wasted, but because it has not yet been consumed.

**Spreading the meals over time**  In humans, babies are breastfed day and night. Baby barn owls behave in a similar way: their mother frequently feeds them pieces of meat night and day. That is why, at hatching, at least one uneaten prey item is found in most nests (80%), compared with only 15% of nests at fledging. At hatching, there are on average four prey items per nest, but the number of remains can sometimes be ridiculously large. In the USA, up to 189 items were found in various states of decay in a single nest, 136 in a nest in Scotland, 79 in another nest in the United Kingdom, and 60 items in a nest in the Czech Republic.

Once the nestlings can eat without maternal help, their daily food requirement is satisfied with few meals, on average 3.4 voles in Switzerland. Because of digestive constraints, nestlings cannot consume these 3–4 voles consecutively during the relatively short period of time when parents deliver food, which explains why many prey items are often found uneaten. Although this situation prevails in many raptors, it can be quite extreme in the barn owl. Fifty per cent of the food is delivered in the first 2.5 hours of the night, but the nestlings spread their meals throughout the night. If some prey items are not eaten by sunrise, they are progressively consumed before the evening.

## FUTURE RESEARCH
- It is not only nestlings that consume prey remains lying in the nest; their parents do as well. How prey remains are shared within the family should be investigated.
- The time when each nestling consumes food should be recorded. Indeed, each individual eats at different times of the day and night, implying that all siblings do not simultaneously compete to monopolize the food delivered by the parents. This could be to reduce sibling competition.

## FURTHER READING
Baudvin, H. 1980. Les surplus de proies au site de nid chez la chouette effraie, *Tyto alba*. *Nos Oiseaux* **35**: 232–238.
Kaufman, D. W. 1973. Captive barn owls stockpile prey. *J. Field Ornithol.* **44**: 225.
Roulin, A. 2004. The function of food stores in bird nests: observations and experiments in the barn owl *Tyto alba*. *Ardea* **92**: 69–78.

An unrelated juvenile has joined a family of four younger nestlings. Perhaps this individual flew to join the family to opportunistically steal food delivered by the parents that was meant for their four offspring.

## 9.5  ADOPTION

# Romulus and Remus

In a variety of organisms, parents sometimes care for unrelated offspring. For example, human newborns can be breastfed by other women (e.g. wet nurses), and adoption is common. In altricial birds, adoption is rare because nestlings cannot reach the nest of another family before they are able to fly. In the barn owl, however, fledglings parasitize care from foster parents, probably more frequently than the few reported observations suggest.

In precocial birds, such as ducks and gulls, nestlings can visit other families by walking or swimming, whereas in altricial birds, such as storks and raptors, abandoning the natal nest to search for a foster family requires the ability to fly. Why do foster parents feed unrelated offspring, and why do fledglings visit other nests?

**Parents accept adoptees**  Barn owl populations are generally rather thinly scattered, implying that recently fledged owls have few opportunities to commute between nests. Hence, adopting a foreign owl that just recently acquired the ability to fly has been reported once in the Netherlands, once in the USA (in Utah), and four times in Germany. In Switzerland, the monitoring of 257 nests twice during fledging revealed seven cases of fledglings roosting during the day in a nest different from the one in which they were born and raised. One of these two-month-old fledglings was in a nest containing one-month-old nestlings; it left the nest approximately 30 minutes before nightfall – was this perhaps because the foster parents would have detected it, given its advanced age? To remain unnoticed, a fledgling should therefore invade families with new nest-mates of similar age, which is what is usually observed. This suggests that kin recognition based on plumage traits or vocalizations has not evolved. Why is this? A possible reason is that adoption is too rare to promote the evolution of complex kin recognition mechanisms.

Adoptions occur not only when parents adopt foreign nestlings but also when they adopt foreign eggs – and they might not even be barn owl eggs. In Europe, when a barn owl pair takes over a cavity occupied by kestrels, the female owl is likely to incubate her eggs along with the kestrel eggs. She does not neglect the kestrel eggs, probably because of the risk of confounding them with her own eggs. When all eggs hatch, barn owl parents can sometimes raise owlets and kestrels successfully!

**Parasitism?** Fledglings can invade other nests looking for extra help from foster parents, such as when their parents do not feed them appropriately or are dead. This could explain why nest-switchers as young as 64 days of age are found in other nests as close as a few hundreds metres from their natal nest. At that age, the nestlings are not yet independent from their biological parents. The observation of an injured 89-day-old fledgling found in a foster nest confirms the idea that some individuals may desperately search for extra parental care.

Juveniles which have already broken ties with their biological parents can have difficulty hunting in the early days of independence and try to prolong the period of parental care. However, nest-switchers can be really 'old' (up to 144 days) and far from their nest (up to 104 kilometres), suggesting that fledglings can still look for extra parental care despite already being independent from their biological parents. Given this information, juveniles could also opportunistically steal food from other families regardless of whether they are independent from their parents and need extra parental care.

**Sexually selected infanticide and adoption** The above examples describe situations in which young individuals invade a family. There are also cases of a non-breeding male killing or adopting nestlings in order to mate with the mother. In Switzerland, two males were suspected to have killed a brood to reproduce with the female. Having failed to produce any fledglings, the mother may have no other choice than to start a replacement clutch with the killer, as is frequently observed in lions.

Fortunately, however, securing a mate can involve more peaceful behaviour. The death of a male barn owl in Germany should have sealed the fate of his five-week-old offspring. Luckily, a new male adopted the chicks, raised them and started a clutch with the mother in the same nest.

## FUTURE RESEARCH
- Kin recognition should be tested to find out whether parents and nestlings can identify foreign fledglings. In particular, it should be determined whether the ability to recognize a foreign fledgling depends on the age difference between the foster and biological offspring.
- GPS devices will allow researchers to monitor the activity of recently fledged birds over long periods of time. This will provide information about the frequency and duration of adoption and the importance of these factors for fledgling survival. If fledglings visit other nests mainly at night, the use of GPS will allow ornithologists to correctly monitor the frequency of nest-switching. So far, all adoption cases have been recorded during daylight hours, yet most nest-switchers may invade foster families at night to gather some prey items before returning to their natal nest to sleep during the daylight hours.
- Do nest-switchers behave in a specific way to successfully invade a foster family? For instance, they may avoid calling to prevent being recognized as a stranger.
- Nest-switching could be a way to find extra food resources, shelter or protection against predators. However, these functions have not yet been clearly demonstrated.

## FURTHER READING
Charter, M., Izhaki, I. and Roulin, A. 2018. The presence of kleptoparasitic fledglings is associated with a reduced breeding success in the host family in the barn owl. *J. Avian Biol.* **49**: e01770.

Roulin, A. 1999. Natural and experimental nest-switching in barn owl *Tyto alba* fledglings. *Ardea* **87**: 237–245.

# 10 Sibling interactions

**10.1** Timing of nestling activities
**10.2** Sibling negotiation
**10.3** Begging behaviour
**10.4** Stealing food from siblings
**10.5** Food sharing among siblings
**10.6** Mutual preening
**10.7** Social huddling

A young barn owl
flapping its wings.

## 10.1 TIMING OF NESTLING ACTIVITIES

# Never sleep

The barn owl is nocturnal, and so should be the nestlings. However, monitoring of nestling behaviour and cerebral activity over 24 hours has revealed a completely different pattern than initially expected. Nestlings sleep as much at night as during the day, with some behaviours (especially the action of preening siblings) being even more frequent during the day than at night and others showing two peaks of activity, at sunset and sunrise.

What do nestlings do during their long days? Do they eat at night and sleep during the day, as one would expect from a nocturnal bird, or do they display a more complex set of activities?

**Nestlings are not strictly nocturnal** Across most of their range, adult barn owls bring all the food to their nestlings at night and are strict nocturnal hunters. Do nestlings align their activity to that of their parents? They do in part, as they vocally compete for food resources (*sibling negotiation* in the graph opposite) and eat (*nestling feeding*) mainly at night, and these activities decrease from sunset to sunrise, as does parental food delivery (*parental feeding*). However, they move (*nestling locomotion*), preen their plumage (*self-preening*), peck their siblings (*pecking*) and flap their wings (*wing flapping*) only slightly more often at night than during the day, and they even preen their siblings (*allopreening*) more frequently during the day. To be able to be so active round the clock, nestlings should sleep not only during the day, as expected from a nocturnal animal, but also at night – and indeed, sleep (*REM sleep* and *non-REM sleep*) is divided roughly equally between day and night. Nestlings show two peaks of activity around sunrise and sunset with respect to locomotion, flapping their wings during cerebral wakefulness, and to a lesser extent in allopreening and pecking their siblings.

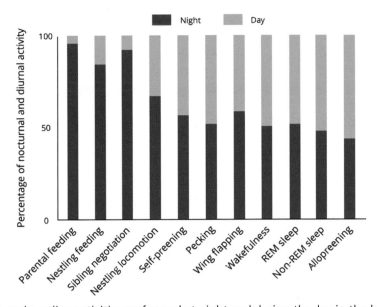

Parental feeding and nestling activities performed at night and during the day in the barn owl in Switzerland. Parents bring food at night, which nestlings consume mainly at night. Not surprisingly, nestlings vocally negotiate access to food resources mainly at night, although this activity starts before it is dark. Although nestlings move and self-preen more often at night than during the day, they frequently perform these activities during the day. Non-REM sleep and allopreening of siblings are performed slightly more often during the day than at night. Nestlings are cerebrally awake, perform REM sleep (i.e. deep sleep), peck their siblings and flap their wings as often at night as they do during the day. From Scriba et al. 2017.

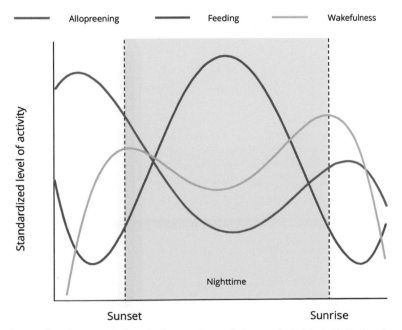

Activity in nestling barn owls in relation to time of day and night. Activity levels were standardized to compare the different behaviours. 'Allopreening' refers to nestlings using their bills to preen siblings, 'feeding' to nestlings consuming prey items, and 'wakefulness' to the brain being cerebrally awake (i.e. the birds are not asleep). From Scriba et al. 2017.

**Sleep architecture**   Whereas humans have a few long sleeping bouts at night, in birds sleep is partitioned into more frequent and shorter bouts, as measured in barn owl nestlings using non-invasive electroencephalography. The mean duration of wakefulness bouts across the whole 24 hours is 50.7 seconds, that of non-REM sleep is 17.1 seconds, and that of REM sleep (deep sleep) is 12.0 seconds. Sleep might be broken up in this way in order to increase the probability of detecting a predator.

## FUTURE RESEARCH
• The adaptive function of nestlings being so active during the daylight hours still requires research.

## FURTHER READING

Scriba, M., Ducrest, A.-L., Henry, I., Vyssotski, A. L., Rattenborg, N. and Roulin, A. 2013. Linking melanism to brain development: expression of a melanism-related gene in barn owl feather follicles predicts sleep ontogeny. *Front. Zool.* **10**: 42.

Scriba, M., Dreiss, A. N., Henry, I., Béziers, P., Ruppli, C., Ifrid, E., Ducouret, P., Da Silva, A., des Monstiers, B., Vyssotski, A. L., Rattenborg, N. C. and Roulin, A. 2017. Nocturnal, diurnal and bimodal patterns of locomotion, sibling interactions and sleep in nestling barn owls. *J. Ornithol.* **158**: 1001–1012.

Barn owl siblings vocally negotiating priority access to the next prey item delivered by a parent.

## 10.2 SIBLING NEGOTIATION

# Diplomacy

The barn owl is quite unique in the range of social interactions displayed between family members. In a quest to monopolize prey items delivered by the parents, young siblings beg from their parents and compete for food, like all other animals with parental care. Remarkably, barn owl nestlings also negotiate vocally with each other, peacefully determining who has priority access to the next delivered indivisible prey item.

Before 1974, ecologists thought that family interactions were harmonious; since then, however, researchers have agreed that these interactions are often conflictual. However, the idea of harmony is now re-emerging: although young siblings compete over parental resources, they can also cooperate. Paradigms come and go!

### All conditions are met for sibling negotiation to evolve in the barn

owl   A wise Arabic proverb states '*I against my brothers, my brothers and I against my cousins and I, my brothers and my cousins against the strangers.*' Similarly, in nature, young siblings still dependent on their parents sometimes cooperate, ensuring the spread of shared genes. However, empirical evidence of cooperation between young siblings is limited to a few species because the conditions that promote the evolution of cooperation are indeed rather restrictive.

In the barn owl, young siblings vocally negotiate priority access to parental food resources instead of fighting endlessly. As far as is known, this is quite unique in animals. For this form of cooperation to emerge, six conditions must be met:

- **Relatedness.** Animals can be more generous with closely related individuals than with strangers. Therefore, negotiating rather than fighting over sharing a resource is more likely to occur between full siblings than between half-siblings or unrelated individuals. In the barn owl, nestlings are commonly full siblings; therefore, the incentive to negotiate with nest-mates can be high.
- **Costly sibling competition.** When the competition between siblings over parental resources becomes tiring and entails the risk of injury, as is likely in raptors with their sharp bills and claws, it might be better to settle contests peacefully.
- **Predictable contest outcome.** Why not reach an agreement when facing an individual that is much hungrier than you and ready to fight? Individual barn owls consume only three or four prey items per night, implying that once a nestling has consumed a meal, it is momentarily satiated and unlikely to monopolize the next delivered food item. This individual would be likely to allow one of its hungry siblings to consume the next food item.
- **Contested food is valuable.** Usually, negotiation occurs when the coveted resource to be shared is precious. Because barn owl nestlings consume few meals per day, a single meal composed of a small mammal is more valuable than, for instance, one of the hundreds of seeds or invertebrates that passerines consume daily. Negotiation is therefore more likely to occur in the barn owl, whose parents bring approximately twenty prey items per night, than in small passerines, whose parents feed their offspring hundreds of times per day.
- **Resource divisibility.** If several individuals can easily get a slice of a cake, it may be challenging to monopolize the entire cake. At each visit, barn owl parents bring one food item that a single offspring consumes. It follows that only this individual is rewarded for the effort invested to out-compete its siblings and get this item. Therefore, if it is difficult to monopolize a food resource, particularly an indivisible resource like a small mammal, it might prove useless to compete with a highly motivated sibling who is likely to fight intensely for the item. Instead of fighting for hours for no return, it is better to agree that the hungriest individual eats the next food item to be delivered, and that the others wait their turn.
- **Cheap sibling negotiation.** Politicians negotiate ceasefires when war becomes too expensive. Similarly, in the barn owl, negotiating is valuable provided it does not outweigh the costs of competition. Therefore, sibling negotiation should not attract predators and should be energetically cheap, at least cheaper than physically competing.

## Sibling negotiation

A food-satiated individual will not compete for indivisible food resources that it is unlikely to acquire, while a hungry individual will vocally confirm its willingness to fight for these resources. Being informed about each others' need for food, siblings finely adjust investment in competition by momentarily withdrawing from a competition that they are unlikely to win. The hungriest nestling negotiates intensely and in turn begs intensely for food from its parents, while the less hungry siblings are less vocal and refrain from competing and begging. This leaves the door wide open for the most vocal individual to acquire a food item without having to compete too intensely. The negotiation process takes time because siblings challenge each other to determine which individual is the hungriest. Identifying the winner of a negotiation process requires lengthy vocal interactions.

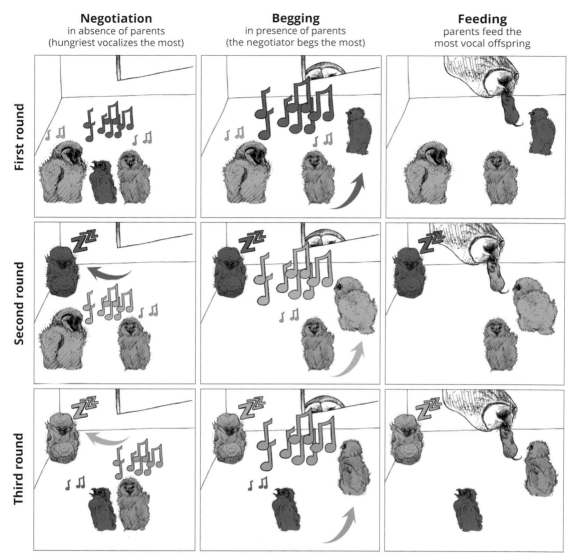

Schematic view of sibling interactions before food delivery and when a parent brings a first indivisible prey item eaten by the orange nestling (*first round*), a second prey item eaten by the grey nestling (*second round*) and a third prey item eaten by the blue nestling (*third round*). Before a parent delivers the first item of food, siblings vocally negotiate, with the orange nestling being more vocal than its siblings; after having listened to their voluble orange sibling, the grey and blue individuals momentarily withdraw from the contest and refrain from begging for food, whereas the orange owlet begs intensely towards its parent, which, in consequence, feeds it. Before the second parental feeding visit, the grey nestling negotiates intensely and, in turn, begs intensely, and before the third parental feeding visit, the blue nestling negotiates and begs intensely. Therefore, sibling negotiation influences the extent to which each nestling begs, and the intensity of begging behaviour determines which individual receives food from a parent.

Even in species such as the barn owl in which sibling negotiation occurs, there might be situations when negotiation does not bring added benefits. A nestling may be willing to allow a sibling to eat first if its parents rapidly return with food. If food supply is poor, a nestling might be less inclined to negotiate, because parents may not deliver another prey item until the following night. Indeed, such situations call for competing fiercely for scarce food items rather than being nice to your brothers and sisters. Stress has a similar effect: when nestlings feel stressed, as measured by an increase in the blood-circulating stress hormone corticosterone, they are less keen to negotiate and compete physically instead.

## The sibling negotiation process

Does negotiation always require verbal communication, as in humans? This is so in the barn owl, in which nestlings produce harsh calls that have no other meaning than to signal how hungry they are in the hope that siblings will withdraw from the competition over the next prey item delivered by a parent. This process, referred to as 'sibling negotiation', occurs in the nest during a prolonged absence of the parents, who are hunting and often far from the nest. When a nestling in a brood is very hungry, it vocally informs its siblings about its motivation to compete intensely to monopolize the next prey item. Being aware of the difficulty in securing this item, its siblings momentarily withdraw from the competition, waiting until this hungry individual has eaten before competing for the next food item. Once the parents return with food, the offspring beg by shouting at the parents. The hungriest nestling, who negotiated most intensely in the absence of the parents by emitting more and longer calls than its less hungry nest-mates, begs more intensely than its siblings. With this behaviour, a nestling increases its chances of being fed by up to 70%, because the parents preferentially transfer food to the most vocal offspring. Once satiated, this individual becomes silent, and its unsuccessful siblings, who were less vocal during the previous period of negotiation and refrained from begging, resume intense vocal negotiation. This is a good strategy: the unsuccessful individuals are now more likely to be fed at the next parental visit, because their previously hungry sibling has already eaten.

## The strongest negotiate and the weakest compete

Although the oldest and largest barn owl nestlings have all the advantages, and are strong enough to take what they want, they negotiate to finely adjust how and when it is best to deploy their superiority. Weaker and smaller nestlings cannot physically outcompete siblings that hatched several days before they did. The alternative for them is to invest more energy in vocalizing and listening to the behavioural response of their dominant siblings to obtain 'permission' to eat the next delivered food item. This works because the older nestlings adjust their behaviour in relation to how siblings negotiate. For example, in response to the emission of playback calls, older nestlings decrease their call rate more than their younger siblings and are also keener to delay feeding. Younger siblings are usually hungrier than their older siblings and hence less inclined to withdraw from the battle to monopolize a food item. This is why the last-born nestlings of a brood strongly display how hungry they are by vocalizing regardless of whether their older siblings are also hungry.

## Negotiation takes time

The likelihood that a nestling monopolizes a food item is higher if it previously vocalized intensely. Thus, why do nestlings not all try to call as intensely as possible all the time? It is probably slightly tiring, and is unlikely to be rewarding because the hungriest nestling of the brood will surely compete intensely once the parents return to the nest with food. Furthermore, emitting a few calls is not enough to clearly demonstrate motivation to compete, which explains why siblings take time to vocally interact while their parents are foraging, to determine who will have priority access to food. The individual that succeeds in imposing itself during the negotiation process is the one that vocalizes the most by producing long monologues. Because it can take a long time to work out which individual is the hungriest, each nestling tries to produce long monologues.

Each night, a nestling emits up to 4600 calls, half of them in the form of short to long monologues; in one instance, an individual produced 1591 calls over the course of two hours without being interrupted by any of its siblings. The other half of the calls are emitted in response to sibling vocalizations, as in a dialogue. To take the lead and produce a monologue, an individual must decide when it should interrupt its vocal sibling and how it can then dominate the vocal interaction. To do so, nestling barn owls use specific turn-taking rules that determine who will take the floor, much like humans. How does a nestling decide when to interrupt a sibling who is calling in a long monologue? Moreover, how does this individual then succeed in turning the situation to its advantage by dominating the floor?

To take the lead, the first call that a challenger emits is soft, as if it were shyly attempting to interrupt its sibling's discourse. While vocally interacting, the challenger replies progressively faster to its sibling, emitting insistent (i.e. longer) calls to silence it. What should the challenger then do after it has successfully silenced its sibling? Continue to vocalize intensely to keep its sibling silent, or stop negotiating at the risk of being considered not as hungry as it initially pretended? The challenger slightly reduces investment in negotiation to evaluate whether its opponent has really given up the

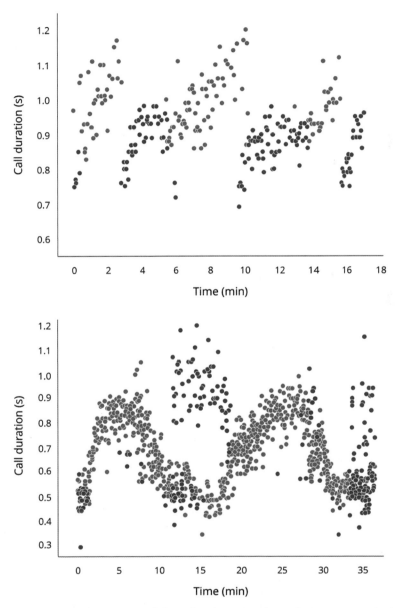

Two barn owl siblings (one in blue and the other in orange) vocally interact. Each dot represents a negotiation call produced during the long absence of the parents, who are foraging. The upper panel shows that siblings do not interrupt each other, with the two individuals vocalizing one after the other. The lower panel shows that siblings use specific turn-taking rules, i.e. the blue individual resumes vocalizations mainly after the orange sibling starts to produce shorter calls.

competition and does not try to call again. If the rival does not give up, it resumes negotiation, and the two pretenders continue to interact vocally, matching the duration of their negotiation calls to each other. These vocal challenges help the siblings to assess each others' relative hunger levels, a necessary condition to finely adjust the rate at which everyone vocalizes. These alternating periods during which one individual vocally dominates the interaction can be time-consuming, particularly if the rivals are similarly hungry. Being an efficient diplomat requires time, dedication and persuasion!

An individual pretending to be needy must be needy. Negotiation is based on trust not only among diplomats but also among barn owls. A single negotiation call that is energetically negligible cannot reflect the degree of hunger. Young barn owls therefore need to repeatedly display their motivation if they wish to deter their siblings from continuing to engage in the current competition. Therefore, negotiating takes time and often starts during the daylight hours, long before the first possible parental feeding visit. However, how does it work? Do nestlings count the calls emitted by a sibling between two parental visits to assess how hungry it is? This would be useless, because many events may intervene over this time, including the consumption of prey items. Instead, information about food need is based on the calls heard just in the previous few minutes. Because the offspring cannot predict when a parent will arrive with food, siblings must negotiate continuously. They need to keep hammering away at it, winning the argument by sheer persistence!

**Barn owls are polite**   How unpleasant it is when a person interrupts us! The same is true for barn owls, whose nestlings actively avoid calling simultaneously with siblings, five times less often than expected by chance. Overlapping the calls of a sibling is counterproductive because nestlings need to hear each other to transfer and acquire reliable information about their food needs. If siblings call simultaneously, they need to produce more and longer calls to make sure that the information is well understood. But might an individual perhaps deliberately overlap the calls of a sibling to blur its vocal signals? Apparently not, because the propensity of siblings to call simultaneously is not related to hunger level and social dominance hierarchy. In fact, calling simultaneously is socially punishable. Playback experiments have shown that if an individual interrupts a sibling, the latter responds by calling more intensely, which silences the rude owlet.

**No need to shout**   Nestling barn owls transfer information about hunger level by modulating the number and duration of negotiation calls. Do they also alter the loudness of their calls? To convey a vocal message, it might be more convincing to shout than to whisper. However, this is not the case, as it has been shown that vocalizing loudly does not improve the resolution of the negotiation process. Given that owlets communicate in a confined nest cavity, there is no need to call loudly in order to convince nearby siblings to withdraw from the competition. Calling loudly may not be a good idea, as it could attract predators.

**Spying and individual recognition**   Like military intelligence, barn owls collect information on the dominance status and food needs of conspecifics before vocally interacting with them. To do so, nestlings eavesdrop on how two given siblings vocally interact, and based on this information they modify their behaviour once confronted by one of those two siblings. This requires the ability to recognize nest-mates based on vocal cues – and indeed, individual barn owl nestlings are easily recognizable from their voices, even by humans. This recognition is also important to ensure that sibling negotiation honestly signals hunger level. A vocally negotiating individual should make sure that the calls it emits are all recognized as its own calls and not mistaken for those of its siblings. Otherwise, siblings may wrongly consider that the calling individual is less hungry than it is. An individual recognition mechanism based on vocal cues could also help nest-mates to verify that the individual who obtains a food item is the one who previously negotiated at the highest level.

**Counting the competitors** Which is better, fighting against many mildly motivated opponents or against a single highly motivated rival? This complex issue was examined by having a nestling listen to playback calls of one, two or four distinct individuals broadcast at different rates. The nestling felt more confident entering into vocal competition when hearing fewer calls produced by few rather than many rivals. When faced with many rivals, the competition was perceived as too harsh, and the owlet withdrew. Nestling barn owls can thus distinguish calls emitted by different individuals, and can use this information to count the number of individuals who are calling.

## FUTURE RESEARCH

- Vocalization behaviour has been studied in nestlings that are able to eat without maternal help. Nothing is known about vocal interactions between siblings, or between parents and offspring, at an earlier age.
- Does sibling negotiation occur in all Tytonidae? For example, in species breeding on the ground, such as grass owls, is the risk of predation too high for loud vocal negotiation to evolve? Additionally, it is still unclear how widespread sibling negotiation is among animals, and birds in particular.

## FURTHER READING

Dreiss, A. N., Ruppli, C. A., Faller, C. and Roulin, A. 2013. Big Brother is watching you: eavesdropping to resolve family conflicts. *Behav. Ecol.* **24**: 717–722.

Dreiss, A. N., Ruppli, C. A., Oberli, F., Antoniazza, S., Henry, H. and Roulin, A. 2013. Barn owls do not interrupt their siblings. *Anim. Behav.* **86**: 119–126.

Dreiss, A. N., Ruppli, C. A. and Roulin, A. 2014. Individual vocal signatures in barn owl nestlings: does individual recognition reinforce the honesty of vocal signalling? *J. Evol. Biol.* **27**: 63–75.

Dreiss, A. N., Ruppli, C. A., Antille, S. and Roulin, A. 2015. Information retention during competitive interactions: siblings need to constantly repeat vocal displays. *Evol. Biol.* **42**: 63–74.

Dreiss, A. N., Ruppli, C. A., Faller, C. and Roulin, A. 2015. Social rules govern vocal competition in the barn owl. *Anim. Behav.* **102**: 95–107.

Dreiss, A. N., Ducouret, P., Ruppli, C. A., Rossier, V., Hernandez, L., Falourd, X., Marmoli, P., Cazau, D., Lissek, H. and Roulin, A. 2017. No need to shout: effect of signal loudness on sibling communication in barn owls *Tyto alba. Ethology* **123**: 419–424.

Dreiss, A. N., Ruppli, C. A., Delarbre, A., Faller, C. and Roulin, A. 2017. Responsiveness to siblings' need increases with age in vocally negotiating barn owl nestlings. *Behav. Ecol. Sociobiol.* **71**: 109.

Johnstone, R. A. and Roulin, A. 2003. Sibling negotiation. *Behav. Ecol.* **14**: 780–786.

Roulin, A. 2002. The sibling negotiation hypothesis. In Wright, J. and Leonard, M. (eds) *The Evolution of Begging: Competition, Cooperation and Communication*, Dordrecht: Kluwer Academic Press, pp. 107–127.

Roulin, A. and Dreiss, A. N. 2012. Sibling competition and cooperation over parental care. In Royle, N., Kölliker, M. and Smiseth, P. (eds) *The Evolution of Parental Care*. Oxford: Oxford University Press, pp. 133–147.

Ruppli, C. A., Dreiss, A. N. and Roulin, A. 2013. Nestling barn owls assess short-term variation in the amount of vocally competing siblings. *Anim. Cogn.* **16**: 993–1000.

Trivers, R. L. 1974. Parent–offspring conflict. *Am. Zool.* **14**: 249–264.

To which offspring should a parent give a prey item? Commonly, the most vocal nestling is fed first.

## 10.3 BEGGING BEHAVIOUR

# Stand up and shout

**Barn owl parents assess offspring begging behaviour to allocate food to the hungriest nestlings, rather than adjusting the feeding rate in relation to short-term variation in offspring food need, as do most other birds. Even if nestlings have recently eaten, they continue to beg from their parents conspicuously, as if they were still hungry. This might motivate parents to keep the feeding rate high over several nights.**

Offspring communicate with their parents to request food, protection and care. Towards the end of the incubation period, young barn owls already start to produce chittering calls inside the egg. Similar calls are emitted during the first week post-hatching in reaction to cold, physical contact and hunger. In the first days of life, the mother reacts to offspring chittering calls 85% of the time, and stops reacting when the offspring are 18 days old, an age when nestlings start to regulate their own body temperature. In the meantime, nestlings progressively produce harsher negotiation and begging calls to display how hungry they are to their siblings and parents, respectively.

**Begging for food** Like most birds, nestling barn owls solicit parental attention to obtain food. They start to beg loudly before their siblings, jostle for the best position in the nest, extend their neck to intercept parents, and flap their wings to disturb siblings. These extravagant behaviours are more strongly displayed by hungry rather than food-satiated offspring, which explains why parents feed the most conspicuous nestlings first. However, the barn owl is quite remarkable among birds.

In species in which the parents feed several chicks at each nest visit, such as in granivorous birds that bring several seeds per visit, nestlings typically escalate begging behaviour. If one individual begs conspicuously, its siblings will try to outperform it to get the extra seeds. This is a never-ending process like an arms race. In the barn owl, this situation is reversed, with nestlings refraining from (instead of escalating) begging if siblings have negotiated intensely beforehand.

Although the youngest individuals of a brood are usually hungrier than older siblings and hence more vocal, older siblings lead the way, and younger siblings follow. The best predictor of which offspring is fed is the extent to which older siblings vocalize in front of their parents, while the begging and the position of younger siblings inside the nest cavity (i.e. being close to or far from the nest entrance where parents arrive with food) poorly predict the outcome of sibling competition. Given their poor resource-holding potential, younger siblings cannot physically outcompete their older siblings. For this reason, before trying to monopolize a food item, younger chicks determine whether their older siblings are willing to concede the impending food item. To this end, younger siblings vocally negotiate and, based on the acquired information, beg loudly only if they have a non-negligible chance of monopolizing the food, i.e. when older siblings refrain from negotiating and then from begging. Older siblings could easily deploy full physical force to obtain what they desire, but such aggressive behaviours are exhausting and potentially dangerous. Therefore, regardless of the younger siblings' need for food, when older siblings are hungry, they negotiate and beg intensely to deter their younger siblings from competing. Even the strongest benefit from negotiating.

## Parental feeding and within-brood food allocation
Although barn owl parents do not adjust feeding rate in relation to short-term variation in offspring hunger levels (i.e. if nestlings beg more loudly, parents do not increase feeding rate, at least not rapidly and not to a large extent), they assess offspring begging to decide to which individual they should give a recently captured prey item. Because parents bring their offspring food at the beginning of each night, prey can accumulate more rapidly than the offspring can consume it. Under these conditions, once nestlings have consumed a few items and are therefore momentarily satiated, should they stop begging? In fact, despite not being hungry, nestlings continue to beg loudly and receive food items from their parents, as observed in Europe and the USA. Although they do not immediately consume the food that is delivered, satiated offspring may pursue intense begging to motivate their parents to keep hunting at a high rate until the end of the rearing period, or they may beg loudly to secure prey items and sit on them for later consumption, preventing siblings from stealing them, as is sometimes observed. Researchers still have difficulty fully understanding the exact function of nestling begging behaviour in the barn owl.

## Preparing for the arrival of the parents
In the quest to monopolize food items, nestlings can either compete against their siblings or attract parental attention, but doing both simultaneously might be difficult. While waiting for their parents, nestlings face the nest entrance in anticipation of a food delivery 80% of the time and look towards their siblings only 40% of the time. In only 9% of observations were nestlings facing their siblings without seeing the nest entrance, while in 43% of observations they could only see the nest entrance and not their siblings. It therefore seems essential for the offspring to communicate face to face with their parents rather than to negotiate face to face with their siblings. This is not surprising, because negotiation calls do not need to be loud to convey information about hunger level and thus do not need to be emitted right besides siblings or in front of them, whereas food is obtained from the parent's bill.

These general patterns vary slightly between younger and older siblings. When positioned near the nest entrance, ready to intercept a parent, younger siblings look more frequently towards the entrance than do their older siblings. This means that when the likelihood of obtaining a food item is relatively high, younger siblings are more vigilant than older siblings in rapidly detecting the

Barn owl nestlings conspicuously
beg for food from their parents.

incoming parent. When positioned at the back of the nest, older siblings look relatively more frequently towards their siblings than do their younger siblings in the same situation. Therefore, when
the likelihood of obtaining a food item is relatively low, older siblings stay alert to see what their
younger siblings are doing.

## FUTURE RESEARCH

- Although nestlings are for the moment not hungry, they continue to beg loudly until the parents have brought all the food required for the night. This may explain why parents do not adjust
feeding rate, or do so only weakly, in relation to short-term variation in offspring need. Parents
may assess offspring begging to bring their daily food requirements night after night rather than
to adjust feeding rate within the same night. In other words, if the offspring reduce (or increase)
begging rate in one night, parents might reduce (or increase) feeding rate over the course of several
nights but not immediately in the same night. To investigate these scenarios, offspring begging
and parental feeding rate should be recorded over multiple nights, to assess whether begging rate
recorded in one night is correlated with feeding rate measured in the following nights.
- When nestlings have recently consumed a food item, they substantially reduce the extent to which
they negotiate with their siblings but not the number of begging calls emitted. It is therefore more
important to finely signal current hunger levels to siblings than to parents. This raises the possibility that if negotiation calls reflect short-term food need (i.e. variation in food need over the course
of a single night), begging may reflect long-term food needs (i.e. over several days). This interpretation should be formally tested.
- The role of sibling negotiation in begging behaviour has been studied, but not the reverse.

## FURTHER READING

Bühler, P. 1980. Die Lautäusserungen der Schleireule (*Tyto alba*). *J. Ornithol.* **121**: 36–70.
Dreiss, A. N., Calcagno, M., Van den Brink, V., Laurent, A., Almasi, B., Jenni, L. and Roulin, A. 2013. The
vigilance components of begging and sibling competition. *J. Avian Biol.* **44**: 359–368.
Godfray, H. C. J. 1995. Signaling of need between parents and young: parent–offspring conflict and sibling
rivalry *Am. Nat.* **146**: 1–24.
Roulin, A. 2001. Food supply differentially affects sibling negotiation and competition in the barn owl
(*Tyto alba*). *Behav. Ecol. Sociobiol.* **49**: 514–519.
Roulin, A. 2004. Effects of hatching asynchrony on sibling negotiation, begging, jostling for position and
within-brood food allocation in the barn owl *Tyto alba*. *Evol. Ecol. Res.* **6**: 1083–1098.

The nestling on the left tries to steal the food item that its sibling has just obtained from a parent.

## 10.4 STEALING FOOD FROM SIBLINGS

# Gentleman thief

**Barn owls are surprisingly peaceful compared to other raptors. Although cannibalizing dead nestlings is frequent, infanticide and siblicide are rare. In nestlings, the only common form of aggressive behaviour is the theft of food items from siblings.**

**Frequency of stealing**  In barn owl nests observed in Switzerland, nestlings steal prey items from their siblings on 3–10% of parental feeding visits. Thefts are perpetrated most often at the beginning of the night when nestlings are hungry. Because younger siblings have more difficulty obtaining food from their parents, they compensate for lack of food by stealing from their older siblings. Pilferage is most frequent when 'cheats' obtain food without having previously vocally negotiated access to it with their siblings: the unsuccessful negotiators re-establish justice by stealing food from these cheats.

**Preventing theft**  The tactics to prevent food stealing are efficient, as only one-third of robbery attempts are successful. Nestlings utter a defensive hiss, shield the food with their wings, turn their back while eating, hide prey items in shaded parts of the nest cavity or even sit on them for later consumption. Quickly swallowing the entire prey item at once is the best tactic used by individuals older than two weeks. Owls can do this surprisingly fast. Once started, it takes only 3.5 minutes to swallow the food, on average, in contrast to the 8 minutes (maximum 19 minutes) required when it is ingested piece by piece. This explains why swallowing prey whole is frequent at the beginning of the night – right after the daily fast and when siblings vocalize intensely as a sign of their level of hunger and willingness to steal a recently acquired food item.

To help digestion, our parents taught us to 'cut up our meat'. Digestibility may explain why older nestlings dismember food items and spread small meals through the night. Their younger siblings less often behave this way, because it is not until they are about 26 days old that they can tear apart pieces of flesh, and furthermore, consuming only parts of prey items entails the risk of the remaining parts being stolen. Starting to consume a prey, leaving it on the side and resuming eating another item is a luxury that only the most competitive nestlings can afford.

Although cannibalism is frequent in the barn owl, siblicide is rare.

## Cannibalism, infanticide and siblicide
Although barn owls furiously defend their nest site, mate and offspring, they are otherwise quiet and peaceful. This is remarkable for a raptor. For instance, wild adults placed in captivity avoid each other or display peaceful behaviour, such as preening each other (allopreening). Of a total of 202 social interactions recorded, only 13 (6.4%) were aggressive, with one owl grasping its opponent with its feet. Young barn owls show similar behaviour. Although nestlings do cannibalize their siblings, death more often occurs from starvation and almost never from siblicide. There are also only a few mentions of parents killing motionless offspring or destroying and eating eggs.

### FUTURE RESEARCH
• Do barn owl chicks prefer some prey species or body parts? To test this, a choice of prey species could be offered to nestlings.

### FURTHER READING
Csermely, D. and Agostini, N. 1993. A note on the social behaviour of rehabilitating wild barn owls (*Tyto alba*). *Ornis Hung.* **3**: 13–22.

Roulin, A., Colliard, C., Russier, F., Fleury, M. and Grandjean, V. 2008. Sib–sib communication and the risk of food theft in the barn owl. *J. Avian Biol.* **39**: 593–598.

Roulin, A., Da Silva, A. and Ruppli, C. A. 2012. Dominant nestlings displaying female-like melanin coloration behave altruistically in the barn owl. *Anim. Behav.* **84**: 1229–1236.

Barn owl nestlings can be generous and feed their younger siblings.

## 10.5 FOOD SHARING AMONG SIBLINGS

# Generosity

**Rare in birds, sharing food between siblings is common in the barn owl. After having consumed a few prey items in the first hours of the night, the older nestlings feed the younger ones, up to four items a night. Food sharing between siblings may release the parents from caring for their younger offspring in the nest, and as a consequence they can spend more time foraging. Additionally, this altruistic behaviour could enhance the survival chances of younger siblings, which indirectly promotes the spread of genes shared with the altruistic older siblings.**

While sharing food between relatives is common in eusocial insects, it is rare in other animals, with the barn owl representing one of the few exceptions. Studying this behaviour in non-human organisms provides relevant information about the evolution of social life and kin selection, with related individuals behaving altruistically.

**Frequency of food sharing** Each night, barn owl nestlings can donate up to four prey items to their siblings, with an average of one prey item shared for every eight deliveries from the parents. A donor nestling walks towards a sibling and drops the prey at its feet or, more rarely, transfers it from bill to bill, in which case the sibling non-aggressively takes the prey from the donor's bill without any reaction. Usually, only the oldest nestlings are generous, particularly at the beginning of the night when the younger siblings are hungry. However, this altruistic behaviour has some limits. Older siblings share food only after having consumed a few prey items themselves.

Paradoxically, food sharing occurs mainly in broods in which the older individuals are super-dominant and obtain most of the food items delivered by their parents, even when they are not hungry. Does this result from older siblings competing to monopolize most of the food, or from parents preferentially allocating food to the oldest offspring? Whatever the reason, older siblings are generous only when the cost of renouncing a prey item is low, i.e. when parents allocate more food to them or when many prey remains are not yet eaten. Older nestlings therefore seem to have some sense of justice, as they redistribute the food among the needy brothers and sisters. This behaviour is innate and pronounced, as incubator-hatched owlets share food even if they never see their parents, and some individuals even try to feed siblings who are already satiated.

**Kin selection** Food sharing evolved in the barn owl possibly to compensate for the frequent absence of the mother from the nest, despite the younger offspring still requiring her help to eat. To feed so many offspring, showing pronounced age differences, the mother would have to participate in foraging and simultaneously take care of the youngest offspring in the nest, two mutually exclusive activities. The mother can also be permanently absent if she has started a second annual brood, sometimes when the youngest offspring of the first brood is only 22 days of age and hence still requires care.

Therefore, from the mother's point of view, support from the oldest chick in taking care of the youngest is most welcome. However, why would the older nestling engage in such behaviour? Why should older siblings share food with their younger siblings who frequently attempt to steal food? Reciprocating good manners could be an explanation, as for example in three-chick broods where nestlings gave more prey items to the sibling who had previously given a food item but not to the other sibling who had not. The condition for generosity therefore seems to be '*I'll feed you now only if you feed me later!*'

Reciprocation cannot be the only reason why siblings share food, however, because it does not explain why older siblings frequently feed younger siblings who steal food rather than share it. One answer is **kin selection**. Older siblings may derive so-called **inclusive fitness benefits** from helping their younger siblings, particularly when they are hungry and vocally solicit food, as has been shown in Switzerland. By feeding their younger siblings, older nestlings can increase the survival prospects of those youngsters, who in turn can spread more of their shared genes: in nests where food sharing was particularly frequent, nestlings had a higher survival rate than in nests where the senior nestlings were less generous.

## FUTURE RESEARCH
- Older nestlings share food with their younger siblings, who receive less care from their mother. Receiving some care from siblings rather than from the mother may have developmental consequences, a proposition that has never been considered in the barn owl.

## FURTHER READING
Falk, J., Wong, J. W. Y., Kölliker, M. and Meunier, J. 2014. Sibling cooperation in earwig families provides insights into the early evolution of social life. *Am. Nat.* **183**: 547–557.

Roulin, A., Da Silva, A. and Ruppli, C. A. 2012. Dominant nestlings displaying female-like melanin coloration behave altruistically in the barn owl. *Anim. Behav.* **84**: 1229–1236.

Roulin, A., des Monstiers, B., Ifrid, E., Da Silva, A., Genzoni, E. and Dreiss, A. N. 2016. Reciprocal preening and food sharing in colour polymorphic nestling barn owls. *J. Evol. Biol.* **29**: 380–394.

Barn owl siblings often preen each other.

## 10.6 MUTUAL PREENING

# I scratch your back, you scratch mine

**Mutual preening, or allopreening, provides mutual benefits to the preener and the preenee. By allopreening parts of another individual's body that are difficult for it to reach, a barn owl not only removes inaccessible ectoparasites but also performs a socially appeasing massage. In the nest, these good manners are reciprocated: a nestling is more likely to allopreen a sibling who previously allopreened it or shared food with it.**

Humans give hugs, chimps scratch each other's back, and barn owls preen their mates and siblings. This tenderness shows how peaceful barn owls can be.

**Mutual preening between adults** Mutual preening (or allopreening) occurs all year long between mated pairs, with the female being more attentive. While allopreening each other, the breeding partners produce 'purring' or 'chirrup' calls to request attention. This is worthwhile, because allopreening can **remove ectoparasites** from body parts that are difficult to reach, such as the head and back. It can also **strengthen social bonds** by providing a relaxing massage. The social function of allopreening seems to be important, as illustrated by owls kept in captivity that are seen to allopreen an individual newly introduced into the aviary.

**Mutual preening between nestlings** Barn owl nestlings engage in complex social interactions, including sibling negotiation, food sharing and allopreening. Nestlings seem to take care of themselves as a priority: they self-preen eight times more often than they allopreen their siblings. Preening oneself seems to elicit allopreening, as an individual is more likely to allopreen a sibling if this individual is already self-preening. Why should a nestling allopreen its siblings instead of resting selfishly or preening itself? The main function of such behaviour may have little or nothing to do with spreading shared genes, because allopreening is unlikely to save siblings. Furthermore, allopreening may not have an important hygienic function, because ectoparasites are rarely located on the head, back and neck, where three-quarters of the allopreening acts are directed. A more likely explanation is that allopreening reduces stress levels by acting as a relaxing massage to reduce the level of circulating corticosterone, the hormone that regulates physiological stress. This is likely to be why younger nestlings allopreen their older siblings more often than the other way around, as a way to soothe them. Thus, the preener and preenee would mutually benefit from allopreening.

Allopreening is a service given to siblings, as a favour for something given in return. A nestling is more likely to allopreen a sibling if it is later allopreened by that individual in return. They can also exchange other commodities, such as allopreening for food, as in primates: at night, a barn owl is more likely to give a prey item to a sibling that allopreened it during the day. This would provide another possible explanation as to why younger nestlings allopreen older siblings, who are more likely than younger ones to share food. Social interactions among barn owl family members are quite fascinating!

## FUTURE RESEARCH
- Allopreening could have non-mutually exclusive functions, including calming individuals, reducing the level of social aggressiveness, helping establish a dominance hierarchy and removing ectoparasites. The relative importance of each of these potential functions should be determined.
- Although owls are expected to be nocturnal, nestlings allopreen their siblings more often during the day than at night. The reason for this is unknown and should be studied.
- The hypothesis that allopreening reduces stress levels should be formally tested. Predictions are that allopreening is more frequent in broods in which social interactions are more stressful and in individuals showing higher levels of circulating corticosterone.

## FURTHER READING
Bühler, P. 1980. Die Lautäusserungen der Schleireule (*Tyto alba*). *J. Ornithol.* **121**: 36–70.
Roulin, A., des Monstiers, B., Ifrid, E., Da Silva, A., Genzoni, E. and Dreiss, A. N. 2016. Reciprocal preening and food sharing in colour polymorphic nestling barn owls. *J. Evol. Biol.* **29**: 380–394.

A crowded nest.

## 10.7 SOCIAL HUDDLING

# Cold as ice

**When the weather becomes cold, animals can huddle close together for warmth. Barn owl nestlings are no exception as they actively huddle, particularly the youngest individuals. The last-hatched nestlings are smaller, have less developed feathers and are in poorer health than their older siblings and consequently have greater difficulty maintaining a stable body temperature. By staying close to their younger siblings, older nestlings release their mother from brooding, which gives her more time to forage.**

Physiological processes are optimized for a narrow range of body temperatures. Maintaining a constant body temperature requires specific physiology and behaviour, particularly in growing birds. Nestlings share the nest, and sometimes the space is so restricted that huddling is unavoidable. In the large barn owl nest cavity, huddling is an active process for the individuals who feel the coldest, who initiate the first body contact by walking towards their siblings.

**Social huddling**  There is a range of ambient temperatures within which body temperature is achieved without changing metabolic heat production (to warm up) or evaporative heat loss (to cool down). This 'zone of thermal comfort' is less restricted in first-born than in last-born nestlings (20–31 °C vs. 21–28 °C) because the last-hatched chicks have less developed feathers and have more difficulties in physiologically regulating body temperature – which is because they eat less than their

When ambient temperatures are low, barn owl siblings huddle against each other.

socially dominant first-hatched siblings. When ambient temperature decreases, they must modify behaviour by leaning against their older siblings to keep body temperature constant.

In endotherms ('warm-blooded' animals that metabolically maintain their body temperature within the zone of thermal comfort), huddling can reduce energy expenditure by anything from 6% to 53%. This is why nestling barn owls spend so much time huddling, 78% (up to 97%) of the time in 24 hours on average. Huddling is particularly prevalent in cold weather, when to prevent a reduction in body temperature, nestlings should both eat more and be brooded by the mother. But how can the mother be simultaneously outside the nest hunting and inside the nest brooding the chicks? A solution to this dilemma is given by the older offspring, who can compensate for the maternal absence by huddling with their younger siblings.

## FUTURE RESEARCH
- When all siblings huddle together, the individuals located in the centre of the brood are warmer. Do siblings compete for this position, or do they rotate, with each individual spending some time in the centre of the huddle? The youngest nestlings, which have more difficulties in physiologically regulating body mass, may spend more time in the centre than their older siblings. These scenarios should be tested.

## FURTHER READING
Dreiss, A. N., Séchaud, R., Béziers, P., Villain, N., Genoud, M., Almasi, B., Jenni, L. and Roulin, A. 2016. Social huddling and physiological thermoregulation are related to melanism in the nocturnal barn owl. *Oecologia* **180**: 371–381.

# 11 Demography

**11.1**   Natal dispersal
**11.2**   Breeding dispersal and migration
**11.3**   Survival prospects
**11.4**   Population dynamics

A barn owl ready to cross the ocean, setting out from Florida in the direction of the Bahamas.

## 11.1 NATAL DISPERSAL

# The grass is always greener on the other side of the fence

**Juvenile barn owls disperse from their natal site to find a new home where food is sufficiently plentiful to raise a family. This journey of a few hundred metres to more than 3000 kilometres is performed soon after fledging. Long before the first reproductive attempt, males have to secure and defend a territory against competitors, which explains why they disperse shorter distances than females.**

Natal dispersal is the process by which juvenile animals leave their natal site to reach the place where they will breed for the first time. They look for a promised land sufficiently far from the birthplace to make it unlikely that they will cross the path of a closely related individual and risk breeding with it. Food should also be plentiful and competition with conspecifics low. Although travelling through unexplored territory is a risky endeavour, it has allowed barn owls to conquer the world.

**Dispersal distance**  Dispersal starts as soon as juveniles are independent from their parents and is mainly performed in the 4–5 months post-fledging, as observed in Germany and Denmark. For example, an individual flew 660 kilometres during the 30 days after it took its very first flight, and another travelled 475 kilometres in the 40 days after fledging. Fast movements are necessary to locate the best foraging spots and for males to defend a nesting place against competitors long before initiating the first breeding attempt. As a consequence, the territorial males move shorter distances than females, as shown in restricted study areas in Europe where, to reach their first breeding site, yearling males and females move on average 9 and 11 kilometres, respectively.

These distances are not necessarily representative of other populations. In North America, ringed juveniles have been recovered dead further from the natal site (the median distance is 36 kilometres, based on 2993 recoveries obtained since 1923) than in Europe (19 kilometres, based on 23 243 recoveries obtained since 1910), and the proportion of ringed birds that are recovered dead more than 1000 kilometres from the birth site is more than double in North America compared to Europe (2.07% vs. 0.83%).

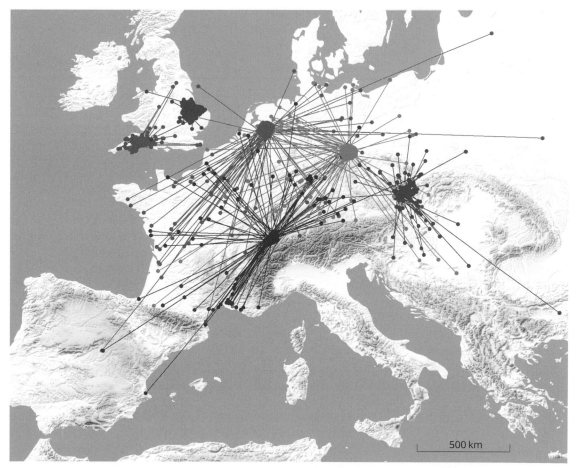

Dispersal movements of European barn owls ringed as nestlings. Data are from a few ringing schemes in the Czech Republic, Hungary, Switzerland, the Netherlands and the United Kingdom. This map shows that barn owls avoid geographic barriers, such as seas and mountains, and barn owls from the United Kingdom move shorter distances than those on the continent. The colours have no meaning other than to distinguish ringing schemes. Courtesy EURING databank.

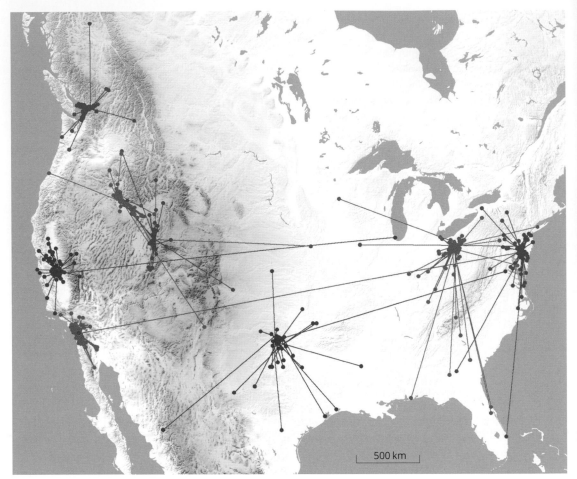

Dispersal movements of North American barn owls ringed as nestlings. Data are from a few ringing schemes in British Columbia (Canada) and a number of states in the USA (California, Idaho, New Jersey, Ohio, Oklahoma and Utah). This map shows that dispersal movements are, on average, longer in North America than in Europe and that owls located on the western side of the Rockies move shorter distances than do owls ringed on the eastern side. The colours have no meaning other than to distinguish ringing schemes. Courtesy USGS Bird Banding (2017). North American bird banding and band encounter data set. Patuxent Wildlife Research Centre, Laurel, MD. 2017/11/01.

**Why not stay at home?** Natal dispersal is a typical behaviour in all bird species. With few exceptions – for instance, three individuals breeding at the site where they were born in Germany, the United Kingdom and the USA – barn owls leave their home to breed. Finding a high-quality foraging territory including an appropriate nest site is one of the most important activities in the first year of life.

But why should a juvenile born in a region where many breeding sites are available take the risk of moving a great distance before settling? Over the last century, ornithologists have added many nest boxes over relatively small areas, so that owls currently do not have to travel far to find a nest site. This may partly explain why, in Europe, dispersal distances have strongly decreased from a mean of 39 kilometres in 1930 to only 4 kilometres in 2015.

It appears that the environmental conditions prevailing during dispersal (i.e. abundance of prey and nesting sites) largely determine how far juveniles disperse, and that whether an individual fledged in poor or prime condition is less important. In most studies, hatching date, hatching rank and brood size were weakly related to dispersal distance. In a few studies, birds born early or late in the season dispersed a greater distance than those born in the middle of the breeding season, individuals from first annual broods dispersed further than those from second annual broods, and individuals from large broods started to disperse earlier than those from smaller broods. Although the environment can affect dispersal behaviour differently between years and populations, the general pattern is that dispersal depends on post-fledging rather than pre-fledging environmental conditions.

Genetic factors may also be at play, as shown in Switzerland where, in their first year of life, young barn owls tend to disperse about the same distance as their same-sex parent. Long trips might be even undertaken as a family. Two siblings from Sweden were recovered in the same village in Denmark, two siblings from Wisconsin at 1920 kilometres in Florida, and two from Alsace-Lorraine at a distance of 700 and 800 kilometres in Brittany; two from the Czech Republic were found dead in the same tank 17 kilometres from their nest 288 days after ringing, and two others were found at the same place 8 kilometres from their birth site 114 days after ringing. The bonds between siblings therefore remain strong long after fledging – but not to the point of breeding together, given the low rate of inbreeding.

**Geographic barriers** Barn owls may not be afraid of the dark, but they do tend to avoid flying over mountains, and only a few have attempted such a flight. One bird has been captured at an altitude of 2000 metres in the Swiss Alps, and 26 birds ringed north of the Alps (Belgium, Czech Republic, Germany, Hungary, Luxembourg, the Netherlands and Switzerland) have been recovered in Italy, suggesting that barn owls can on occasion cross high-altitude mountains.

Even if barn owls usually avoid flying over the sea, they can undertake such journeys – and must have done so in order to colonize remote islands. There are 21 records of barn owls ringed on the European continent crossing the 34-kilometre-wide English Channel to reach Great Britain. In addition, one bird ringed on Bengalis Island in Sumatra crossed the 50-kilometre-wide Straits of Malacca, one owl ringed in Germany was recovered on Gotland island, 130 kilometres from the coast, one owl ringed in New Jersey was found 1250 kilometres across the ocean in Bermuda, and a Tunisian barn owl crossed the Mediterranean to reach southern Italy. Direct observations of owls flying 2 metres above the water 8 kilometres from the Baltic Sea shore, and another seen offshore 56 kilometres from an island in Hawaii, confirm that barn owls can fly over large bodies of water.

## FUTURE RESEARCH
- More studies are needed on the potential role of hatching date, brood size and hatching rank on dispersal, and on whether nestlings from first or second annual broods disperse different distances.
- The dispersal distances should be measured in relation to the availability of nest sites, with the prediction that individuals move shorter distances in regions where many potential nest sites are available.
- Barn owls sometimes fly over large bodies of water, which must be how they colonized many remote islands. One question is whether individuals who risk flying over water are phenotypically different from those that do not take this risk.

## FURTHER READING
Baege, L. 1955. Beachtlicher Zug junger Schleiereulen. *Falke* 2: 213.

Bairlein, F. 1995. Dismigration und Sterblichkeit in Süddeutschland beringter Schleiereulen (*Tyto alba*). *Vogelwarte* 33: 81–108.

Duffy, K. and Kerlinger, P. 1992. Autumn owl migration at Cape May Point, New Jersey. *Wilson Bull.* 103: 312–320.

Huffeldt, N. P., Aggerholm, I. N., Brandtberg, N. H., Jorgensen, J. H., Dichmann, K. and Sunde, P. 2012. Compounding effects on nest-site dispersal of barn owls *Tyto alba*. *Bird Study* 59: 175–181.

Stewart, P. A. 1952. Winter mortality of barn owls in central Ohio. *Wilson Bull.* 64: 164–166.

Van den Brink, V., Dreiss, A. N. and Roulin, A. 2012. Melanin-based colouration predicts natal dispersal in the barn owl *Tyto alba*. *Anim. Behav.* 84: 805–812.

A wintering barn owl in Montana, USA.

## 11.2 BREEDING DISPERSAL AND MIGRATION

# Wanderlust

Barn owl adults are usually sedentary, although some individuals can disperse hundreds of kilometres between breeding seasons to breed at a different nest site. Males defend a nest cavity against competitors and females visit several males before deciding where to breed. Males are thus more site-faithful than females, and move shorter distances when they switch nest sites. In the cold northern parts of the northern USA and southern Canada, barn owls migrate southwards in winter, a behaviour that is not observed in Europe.

Breeding dispersal defines the movement of adults between two successive breeding sites. Because breeders can defend a territory year-round, the distance between two nest sites used in successive years is typically shorter than the distance between birth site and the site where an individual breeds for the first time.

**Dispersal distances** In the barn owl, breeding and natal dispersal are different processes that do not involve the same constraints. This is why the distance travelled by a juvenile between its birth site and its first breeding site does not predict its dispersal behaviour as an adult. Although some owls have a greater inherent tendency than others to disperse, owls become more sedentary with age, probably because old individuals are better able to defend a nest cavity against competitors than young ones – or perhaps simply because they have had more time to discover where the best sites are located.

Nest sites belong to males. Hence, if several sites are available in a relatively small area, a male can easily go from one place to another to breed, regardless of whether he is a faithful partner. In contrast, if a female intends to divorce, she has to move a greater distance within or between years to find a new partner in another territory (although sometimes barn owls are not so territorial, and different pairs can breed close to each other). Males are thus more often site-faithful (in Switzerland, 46% of males breed at the same place in two successive years) than females (33%), and the mean distance between two successive sites is only 1.8 kilometres in males, compared with 3.8 kilometres in females.

In the barn owl, poor reproductive success often induces divorce, but not necessarily a change in nest site. The individual that makes the decision to divorce may thus more often consider that its partner, rather than the quality of the nest site, is the cause of poor reproductive success. Still, food supply surely explains why barn owls switch nest sites and why they sometimes disperse long distances, perhaps in search of better feeding conditions. In Europe, 2–3% of the adults move more than 100 kilometres between consecutive years (up to 844 kilometres). These explorers are presumably among the most competitive individuals, or they would risk being relegated to the poorest nest sites in the host region, where resident owls already know the best sites.

However, home-loving owls seem to be more frequent. In France in the 1970s, from one year to the next, 97% of barn owls re-used the same nests located in churches, and in Utah 96% of breeding owls were site-faithful. This suggests that it may be better for an owl to stay at home even if the environmental conditions become temporarily poor. In regions where ornithologists put up many nest boxes, breeding birds are less site-faithful, as shown in Scotland, where 29% of breeding owls switch nests between years on average. By offering many potential nest sites to barn owls, ornithologists may change the behaviour of the owls.

**Migratory movements** Juveniles and adults undertake longer-distance movements in America than in Europe. American barn owls can indeed be migratory, with approximately three-quarters of birds moving southwards between August and December and northwards between March and April. Even though the species is not migratory in Europe, long-distance movements are more frequent in the direction west/southwest than in the direction east/northeast, whereas short-distance movements can occur in any direction.

## FUTURE RESEARCH
- The factors that trigger adults to change nest site and that influence the distance that they move to find a new site should be studied – in particular, the relative role of food supply, the availability of nest sites and competition over nest cavities with conspecifics and other animals.
- The reason why barn owls are migratory in North America but not in Europe is unclear. Is this because in North America the distribution range includes extremely cold regions, and the winter temperatures at a given latitude in North America are far colder than in Europe?

## FURTHER READING
Van den Brink, V., Dreiss, A. N. and Roulin, A. 2012. Melanin-based colouration predicts natal dispersal in the barn owl *Tyto alba*. *Anim. Behav.* **84**: 805–812.

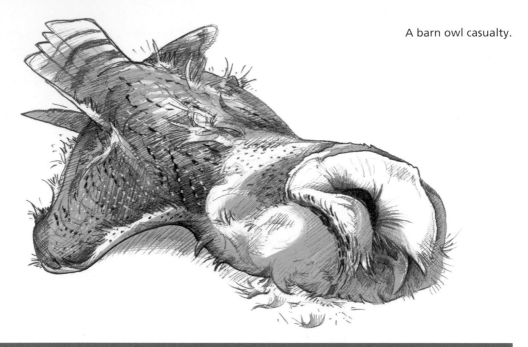

A barn owl casualty.

## 11.3 SURVIVAL PROSPECTS

# Born to lose

**The period during which they gain their independence is particularly dangerous for juveniles, many of which die during this time. Juveniles die from starvation during harsh winters, which is also the main cause of death among adults. However, in most Western countries, deliberate killing by humans was the main cause of death for owls before the 1960s, and road traffic has now become the most frequent cause. The oldest recorded wild barn owl was 23 years of age.**

Ornithologists regularly ring barn owls to obtain data about their dispersal behaviour and risk of mortality. Many aspects of mortality have changed during the last 150 years owing to profound modifications of human–barn owl interactions. Information about mortality comes mainly from barn owls in the northern hemisphere.

**Variation in survival**  In Europe and North America, adults mainly suffer from cold ambient temperatures and reduced food supply, as more adults die in winter (e.g. in the Netherlands, 55% of dead individuals are recovered in the winter) than in any other season (19% in autumn). The effect of harsh winters on survival may be due to the difficulty in hunting small mammals hidden under a thick layer of snow. Fluctuations in vole population size are also an important factor, accounting for 51–73% of the variation in barn owl population size in the Netherlands and Scotland.

   Although juveniles are also sensitive to winter conditions, the period when they gain independence from the parents is the most critical. In the Netherlands, a larger proportion of first-year barn owls die between September and November (44% of all recovered juveniles) than between December and February (33%), even though feeding conditions are worse in winter than in autumn. Individuals born early rather than late in the season therefore have more time to develop foraging skills before feeding conditions decline, and in turn have a greater chance of survival. Because adults and juveniles often die from different causes or at different times of the year, their annual survival rates are usually not correlated, as observed in Switzerland, although exceptionally harsh winters can affect all individuals, regardless of their age.

Reliable estimation of survival probability requires modern capture–recapture methods of analysis to consider the probability that a live barn owl is retrapped by the ornithologists or recovered dead by the general public. According to data about ringed barn owls recovered in Switzerland between 1934 and 2002, the survival probability from one year to the next was 29% for juveniles and 57% for adults. However, these analyses did not take into account the possibility that some birds were not found because they had emigrated outside of the study area. Combined analysis of live-recapture data collected within a restricted study area and dead-recovery data collected within and outside of this area found that annual survival rates for juveniles and adults in Switzerland were 17% and 72%, respectively, in the period 1990–2002.

During the first year of life, survival is therefore dramatically low, and survival during this period in barn owls is lower than in similarly sized long-eared owls and tawny owls. Barn owl life expectancy at 1 year of age is 1.2 years, compared with 1.4 years for the long-eared owl and 2.6 years for the tawny owl, and at 2 years it is 1.8 years, compared with 3.3 and 3.6 years for the other two species. Some lucky barn owls can nevertheless reach the impressive age of 23 years.

**Traffic** Throughout the world, the length and number of roads, and the traffic that uses them, have increased at a worrying pace. In the USA, there are more than 230 million cars driving along more than 6 million kilometres of roads. In the Netherlands, the total length of motorways increased from 350 kilometres to 2000 kilometres in the period 1960–1985, and the number of vehicles increased tenfold, from 450 000 to 4.5 million. In Denmark, cars drove 6.8 billion kilometres in 1958 and 46 billion kilometres in 1999.

Not surprisingly, since the 1960s, road mortality has surpassed hunting as the most common cause of human-induced mortality in a wide range of animals. This is particularly true in the barn owl, which represents more than 30% of all large vertebrates found dead along roads and up to 80% of all dead owls. In the United Kingdom, of all barn owls for which the cause of mortality was known, 6% were victims

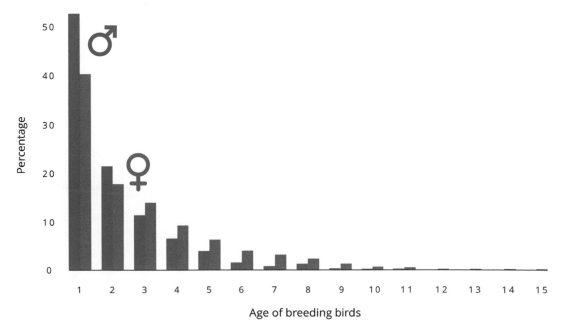

Age of breeding females (*in blue*) and breeding males (*in orange*). This is based on 683 captures of males (346 different individuals) and 550 captures of females (330 different individuals) between 1988 and 2017 in Switzerland. These birds were ringed as nestlings and hence were of known age when recaptured as breeders.

Harsh winter conditions can cause drastic reductions in barn owl populations.

of traffic accidents in 1910–1954, 15% in 1955–1969, 42% in 1963–1989, and 52% in 1982–1986. In the Netherlands before 1963, 65.5% of ringed barn owls died from starvation and 4.4% from traffic; after 1963, these percentages changed to 10.8% and 40.3%, respectively. Similarly, in the Czech Republic, barn owl mortality due to traffic represented 4.3% of all mortalities with known causes in the period 1966–1975, 10.4% in 1976–1985, 22.2% in 1986–1995, and 30.2% in 1996–2007. Traffic mortality is so frequent that juveniles are often killed not far from their birth site, even before they have had a chance to disperse long distances. This might well account for observations that the distance between barn owl birth sites and the locations where owls are found dead strongly decreased in Europe from an average of 47 kilometres in 1930 to 7.5 kilometres in 2015.

In France, approximately 20 000 barn owls die on the roads every year; however, this must be an underestimate, given that other animals displace dead owls, some injured owls may move away to hide, and corpses may not be detected in tall grasses, as reported in Idaho. Several studies carried out in France illustrate the dramatic impact of traffic. Between November 1991 and December 1995, 1731 barn owls were found dead in one study, compared with 811 long-eared owls and 539 common buzzards. This corresponds to one barn owl dying per year per kilometre of road in this specific study! Owls are found dead more often along sections of highway bordered by cereal fields (68%) than along those bordered by meadows (15%) and forests (17%), more often along stretches of road that are higher than the surrounding land than along those that are at a lower level (64% vs. 36%), and more often in sections where common voles are abundant (76%) than in those where common voles are rare (24%).

In Idaho, the barn owl is also the most frequently killed animal along roadsides (32% of all roadside animal deaths), with six barn owls dying every year per 100 kilometres of highway. In this state, mortality is more frequent between October and March than between April and September. In winter but not in summer, highways located in agricultural areas witness more barn owl mortality than those located in shrub–steppe and mixed habitats.

In Portugal, eleven radio-tracked barn owls actively avoided areas close to highways when traffic was dense but tended to fly towards roads if they had streams alongside them or well-vegetated verges, which are havens of peace and tranquillity for numerous prey species. On average, an individual has to cross a highway 111 times before being hit by a vehicle, and this risk is highest along portions of highways that owls usually tend to avoid.

How can we mitigate the dramatic impact of roads? Barn owls fly close to the ground, so they could be forced to fly high above roads by placing artificial structures or trees close to the road. New roads should also be constructed below the surrounding fields. Another mitigation measure is mowing

In Indonesia, as in other countries, animals, barn owls in particular, are captured and sold on the market as pets. This picture was taken in a street animal market in Jakarta in November 2017. The Harry Potter movies have had a large influence on the sale of owls in the bird markets of Indonesia. After watching the movies, children want to have an owl at home. As a result, the capture and sale of these birds have skyrocketed in recent years. © Joan de la Malla

road verges to reduce prey availability and, in turn, predator attraction to these areas. (However, this method might not be beneficial to the environment more generally.) Finally, we should improve the quality of farmland as a foraging area, to prevent owls flying close to roads.

## Other causes of mortality
Deliberate killing of barn owls has strongly decreased over the last century, demonstrating that education has helped people understand that raptors and owls are friends rather than enemies. In the 1980s in France, humans intentionally destroyed only two nests out of more than 1000, and in the United Kingdom, shooting mortality decreased from 12% (of all known causes of mortality) in 1910–1954 to 5% in 1955–1969 and 2% in 1963–1989. Unfortunately, in Spain, humans removing nestlings from their nest or shooting birds was still regularly observed in the 1980s and 1990s. In many countries worldwide, particularly around the Mediterranean, humans still actively kill birds of many species.

Worldwide, approximately 1 million birds are killed annually by wind turbines. In several surveys in Europe, of 19 010 birds found dead below turbines, only four were barn owls. However, in the USA, barn owls may be more susceptible to wind turbines, as a survey between 1998 and 2003 showed that these constructions killed at least 50 barn owls, 213 red-tailed hawks, 70 burrowing owls, 59 American kestrels and 18 great horned owls. Among other causes of mortality, barn owls can become entangled in vegetation, probably because of their velvet plumage; strike windows; become caught in fences or trapped in buildings; and die from disease. Although power lines are dangerous to large birds, such as eagle owls, storks and vultures, few barn owls seem to have been victims of them.

## FUTURE RESEARCH

- Traffic is a major cause of mortality in the barn owl. It is still unclear whether individuals in poor condition take greater risks to forage along roads than do healthy individuals.
- Information about mortality has been obtained mainly from barn owls in the northern hemisphere, where extreme winters have devastating effects on barn owl populations. The impact of drought in arid regions is unknown, as are the main threats that Tytonidae face in the tropics and subtropical regions.

## FURTHER READING

Altwegg, R., Roulin, A., Kestenholz, M. and Jenni, L. 2006. Demographic effects of extreme winter weather in the barn owl. *Oecologia* **149**: 44–51.

Altwegg, R., Schaub, M. and Roulin, A. 2007. Age-specific fitness components and their temporal variation in the barn owl. *Am. Nat.* **169**: 47–61.

Baudvin, H. 1986. La reproduction de la chouette effraie (*Tyto alba*). *Le Jean le Blanc* **25**: 1–125.

Boves, T. J. and Belthoff, J. R. 2012. Roadway mortality of barn owls in Idaho, USA. *J. Wildl. Manag.* **76**: 1381–1392.

Bruijn, O. de. 1994. Population ecology and conservation of the barn owl *Tyto alba* in farmland habitats in Liemers and Achterhoek (the Netherlands). *Ardea* **82**: 1–109.

De Jong, J., van den Burg, A. and Liosi, A. 2018. Determinants of traffic mortality of barn owls (*Tyto alba*) in Friesland, the Netherlands. *Avian Conserv. Ecol.* **13**: article 2.

Forman, R. T. T. and Alexander, L. A. 1998. Roads and their major ecological effects. *Annu. Rev. Ecol. Syst.* **29**: 207–231.

Gomes, L., Grilo, C., Silva, C. and Mira, A. 2009. Identification methods and deterministic factors of owl roadkill hotspot locations in Mediterranean landscapes. *Ecol. Res.* **24**: 355–370.

Grilo, C., Sousa, J., Ascensão, F., Matos, H., Leitão, I., Pinheiro, P., Costa, M., Bernardo, J., Reto, D., Lourenço, R., Santos-Reis, M. and Revilla, E. 2012. Individual spatial response towards roads: implications for mortality risk. *PLoS One* **7**: e43811.

Grilo, C., Reto, D., Filipe, J., Ascensão, F. and Revilla, E. 2014. Understanding the mechanisms behind road effects: linking occurrence with road mortality in owls. *Anim. Conserv.* **17**: 555–564.

Guinard, E., Juillard, R. and Barbraud, C. 2012. Motorways and bird traffic casualties: carcasses surveys and scavenging bias. *Biol. Conserv.* **147**: 40–51.

Newton, I., Wyllie, I. and Asher, A. 1991. Mortality causes in British barn owls *Tyto alba*, with a discussion of aldrin–dieldrin poisoning. *Ibis* **133**: 162–169.

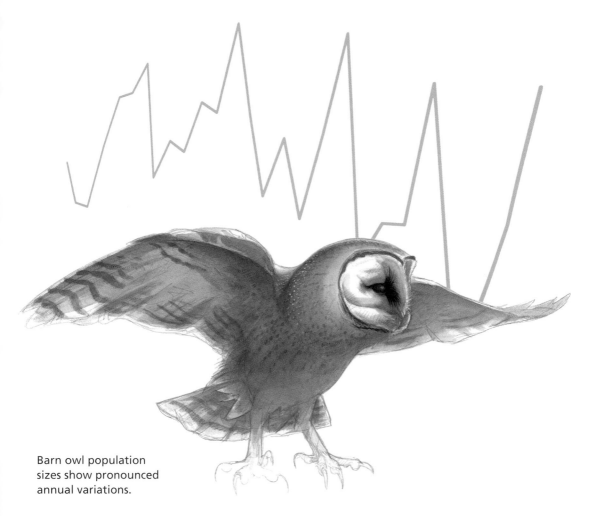

Barn owl population sizes show pronounced annual variations.

## 11.4 POPULATION DYNAMICS

# Replenishing the workforce

**Barn owls commonly show greater annual variation in population size than other raptors of comparable size, such as kestrels. Owl populations can crash after a harsh winter and when small mammals are scarce. When food is in short supply, dense populations decrease in size as a result of pronounced adult mortality. Small populations can increase in size if food is plentiful, which improves the likelihood of a bird being recruited as a breeder in its first year.**

Barn owls face many threats from humans, including persecution, road mortality, pesticides, reduced food supply due to habitat destruction and fragmentation, loss of breeding places and catastrophic weather conditions that may become more frequent with climate change. The high sensitivity to environmental factors, such as harsh winters in Europe and North America, may account for the evolution of certain key aspects of barn owl biology. This includes early sexual maturity, the capacity to produce 2–3 annual broods with many offspring and the capacity to disperse long distances to use ephemeral resources, such as the water that becomes available after intense rain in Australian deserts. These adaptations allow the barn owl to quickly recover from sharp population declines.

**Survival, emigration and immigration** Understanding the mechanisms that drive variation in population size, so-called **population dynamics**, requires detailed knowledge of mortality, emigration and immigration. Small populations can increase if locally born individuals are recruited and immigrants settle from other regions. Once the carrying capacity is reached and an area can hardly support any extra birds, the owl population declines as a result of density-dependent mortality, with many individuals competing over limited resources.

The number of individuals that a region can support is limited by food supply and the availability of potential nest sites. In Switzerland, in years when survival and reproductive success are high, three-quarters of juveniles leave their natal region, probably to avoid intense competition with conspecifics. Only 1% of adults emigrate, but they are not free of competition. In years when more adults have left the study area, those who remain often move to change nest site, which suggests that competition for better nesting sites could be higher in those years. In years when emigration levels are high, even though many individuals are lost to the local population, immigration can be even more pronounced, leading to an overall increase in population size.

In Switzerland, 78% and 62% of breeding females and males, respectively, were born outside of the study area. In a study area in Utah, 77% of individuals were immigrants, while in a study area in Germany, 55% of breeders were immigrants, 14% were locally born new recruits, and the rest (31%) were breeding in the study area in previous years. Population turnover is therefore pronounced; for example, in the Utah population each year, between 25% and 79% of individuals (48% on average) are breeding for the first time. Note, however, that these numbers depend on the size of the study area because locally born owls are more likely to settle outside of a small study area than outside of a large one. Likewise, a small study area will almost inevitably host a larger proportion of immigrants than a larger area.

**78% of females are** immigrants
**62% of males are** immigrants

75% of locally born juveniles emigrate out of the study area

1% of breeding adults emigrate out of the study area

In Switzerland, a large proportion of the barn owl population in a defined area is composed of immigrants, and many of the individuals born inside this area emigrate.

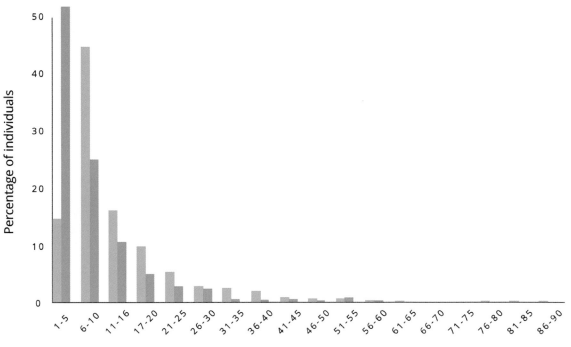

Lifetime reproductive success of breeding barn owls, showing the number of eggs produced in a lifetime in green and the number of fledglings in blue. This is based on data collected from 672 different males and 864 different females from 1990 to 2017 in Switzerland. For example, the category '6–10' indicates that 44.7% of breeding birds produce 6–10 eggs in their lifetime, and 25.0% produce 6–10 fledglings.

## Survival rates make more difference than reproductive success Environmental factors that simultaneously impair the survival and reproduction of all individuals regardless of age strongly increase the likelihood of extinction. Populations are more stable if individuals are not all similarly sensitive to environmental factors, because some of them will survive to replenish their population.

Between 1934 and 2002, the Swiss barn owl population crashed twice during the exceptionally harsh winters of 1952/1953 and 1962/1963, which affected both juvenile and adult survival. When weather conditions are less extreme, population dynamics are mainly sensitive to variation in survival rate, with that of juveniles having the greatest effect, followed by adults older than two years and then by yearlings. Population growth rate is less sensitive to variation in reproductive success, a life-history trait that varies to a much higher degree between years than survival rates. This is the case in Scotland, Switzerland and Utah, where the total number of fledglings produced in one year does not predict population size the next year.

## Lifetime reproductive success Some individuals contribute disproportionately to population growth compared to others. In Utah, where 357 nesting attempts were monitored between 1977 and 1995, some females produced up to seven broods, but the mean number of breeding attempts per individual was only 1.4. As a result, over the course of a lifetime, the most productive female laid six times as many eggs as the population mean (66 vs. 10) and produced more than eight

times as many fledglings (50 vs. 6). In other words, only 12% of breeding females (21 out of 179) had descendants breeding in the local population, with a maximum of seven breeding descendants. In Switzerland between 1990 and 2017, one female laid up to 90 eggs and produced up to 60 fledglings in her lifetime. The maximum number of offspring that were later breeding in the study area was fourteen for one male in his lifetime. The probability of having at least one offspring breeding in a study area is higher if barn owls breed early rather than late in the season.

## FUTURE RESEARCH

- Only one study has examined whether reproductive success declines with age. Senescence in hatching success was detected, with eggs being less likely to hatch when the father was old rather than young. A more thorough study is required to investigate whether reproductive parameters increase from a young age to reach a peak in middle age before senescence occurs.
- In years when survival and reproductive success are high, both emigration and immigration are pronounced. Are emigrants trying to avoid competition with conspecifics? If so, are emigrants less competitive than residents, and are immigrants more competitive than emigrants?

## FURTHER READING

Altwegg, R., Roulin, A., Kestenholz, M. and Jenni, L. 2003. Variation and covariation in survival, dispersal, and population size in barn owls. *J. Anim. Ecol.* **72**: 391–399.

Altwegg, R., Roulin, A., Kestenholz, M. and Jenni, L. 2006. Demographic effects of extreme winter weather in the barn owl. *Oecologia* **149**: 44–51.

Altwegg, R., Schaub, M. and Roulin, A. 2007. Age-specific fitness components and their temporal variation in the barn owl. *Am. Nat.* **169**: 47–61.

Bruijn, O. de. 1994. Population ecology and conservation of the barn owl *Tyto alba* in farmland habitats in Liemers and Achterhoek (the Netherlands). *Ardea* **82**: 1–109.

Kniprath, E. and Kniprath, S. S. 2014. Schleiereule *Tyto alba*: Eigenschaften und Bruterfolg einer zweiten niedersächsischen Population. *EulenWelt* **2014**: 43–65.

Marti, C. D. 1997. Lifetime reproductive success in barn owls near the limit of the species' range. *Auk* **114**: 581–592.

Sæther, B.-E., Grøtan, V., Engen, S., Coulson, T., Grant, P. R., Visser, M. E., Brommer, J. E., Grant, B. R., Gustafsson, L., Hatchwell, B. J., Jerstad, K., Karell, P., Pietiäinen, H., Roulin, A., Røstad, O. W. and Weimerskirch, H. 2016. Demographic routes to variability and regulation in bird populations. *Nature Comm.* **7**: 12001.

Taylor, I. R. 1994. *Barn Owls: Predator–Prey Relationships and Conservation*. Cambridge: Cambridge University Press.

# 12 Plumage polymorphism

12.1   Colour polymorphism
12.2   Genetics of plumage polymorphism
12.3   Sexual dimorphism in plumage traits
12.4   Age-related changes in plumage traits
12.5   Mate choice
12.6   Sexually antagonistic selection
12.7   Adaptive functions of whitish and reddish colouration
12.8   Adaptive functions of small and large black spots
12.9   Adaptive functions of few and many black spots
12.10  Geographic variation in plumage traits

In the ruff, black males are more aggressive, and white males adopt sneaky behaviour to copulate with females.

## 12.1 COLOUR POLYMORPHISM

# Black and white

**In most barn owl populations, no two individuals are similarly coloured. Although the plumage on the underside of the body varies between individuals from white to dark reddish-brown and from spotless to heavily marked with black spots, distinct colour forms, the so-called 'colour morphs', can be defined. Colour polymorphism can be used to understand the evolution and adaptive function of different phenotypes and how this diversity persists.**

The colour differences that can be observed between and within barn owl species are due to melanin pigments. Melanin is the most common pigment in animal integuments and is responsible for some of the most fabulous colour patterns. The grey to black **eumelanin** is found not only in animals but across almost the entire tree of life, while the pale to dark reddish-brown **phaeomelanin** is present mainly in mammals and birds – though evidence accumulates supporting its synthesis also in reptiles and insects.

In the early evolution of life, melanin protected bacteria and fungi against biochemical attack from the environment. In unicellular, invertebrate and vertebrate organisms, melanin still retains the same protective function against ultraviolet radiation and parasites. This pigment also helps accumulate warmth, and some of the genes involved in the production of melanin (melanogenesis) regulate numerous physiological and behavioural functions (they are 'pleiotropic' genes). So many functions for a single molecule! While the yellow to orange carotenoid pigments are extracted from nutrients, at least 120 genes control the endogenous production of melanin, reflecting that it is a key adaptation.

**The study of plumage traits in the barn owl** Studying the adaptive function of colour traits in birds is very popular among evolutionary ecologists. However, for unknown reasons, this interest has rarely crossed from academia to the world of the amateur ornithologist, who usually considers plumage traits of interest only as a means of identifying species, sex or age. Very few researchers have investigated why the plumage of barn owls is so diverse, and what could be the adaptive function of different colour morphs.

I started to be fascinated by this variation when I first handled a pair of barn owls in which the male was white and the female dark reddish. This single observation motivated me to pursue an academic career and dedicate it to the study of plumage colouration and spottiness. At the start of these studies in 1990, friends and other ornithologists laughed at me, thinking that an apparently trivial detail such as the size and number of spots could have no function. In their eyes, I was just wasting my time by measuring and counting spots on barn owls – and in fact I had no answer when they asked me why it was so interesting. After years of intensive research, however, the results are really intriguing – showing connections between plumage traits, physiology, behaviour, personality, ecology and finally natural and sexual selection. It is astonishing to realize that such small plumage details can be so important, especially in a nocturnal bird. These discoveries raise important questions about the genetic mechanisms underlying such links.

**Barn owl plumage traits** Compared to 'typical owls' or 'true owls' (family Strigidae), barn owls and relatives (family Tytonidae) are very distinct, with a cream-coloured back and whitish underparts. Inter-individual variation in colouration, from white to dark reddish-brown and from immaculately spotless to heavily marked with black spots on the feather tips, is surprisingly similar among barn owls, grass owls and masked owls. The only exception is the sooty owl, which is blackish instead of whitish and displays white spots instead of black spots, like a photographic negative of the other Tytonidae.

In rare cases, a barn owl can display a completely different colouration owing to mutations. For instance, some melanic barn owls have been discovered in the USA, Germany and the Czech Republic. This indicates that an entirely black plumage, similar to that of the sooty owl, could easily evolve anywhere on the planet. However, this type of colouration in barn owls may not be as advantageous as in sooty owls, and remains rare.

Colour aberrations are scarce but can still be observed in the wild because they are likely to be often encoded by recessive alleles that must be inherited from both parents to affect the phenotype. In the Czech Republic, two captive siblings produced 23% melanic offspring (seven out of a total of thirty), which is very close to the predicted 25% if melanism is encoded by a recessive allele. Besides melanic forms, there have been only two reports of albinism, one of erythrism (unusual reddish pigmentation) and several cases of partial loss of pigmentation (leucism) in Argentina, Germany, Sardinia and Switzerland.

**Colour polymorphism** In many owl and raptor populations, individuals of the same sex and same age display one of several colour morphs. When morphs are sufficiently common not to be considered aberrations, the species is said to be colour polymorphic. In most barn owl populations across the world, although colouration varies continuously between white and dark reddish-brown, a limited number of morphs can be distinguished, such as the 'white' and 'red' morphs. The expression of colour polymorphism is commonly under the control of just a handful of genes, with each allele encoding a different morph. The study of colour polymorphism can be used to answer three major questions:

(1)  **How can different life forms evolve?** A mutant for a new colour variant can spread in a population if it confers an initial advantage to its bearer; otherwise the new morph would disappear as quickly as it emerged or, at least, remain at a very low frequency. For example, a melanic mutant in an otherwise pale diurnal raptor can be more cryptic in a dense forest and have an advantage over pale conspecifics when hunting in such a habitat (also expanding the species' range of

Plumage traits are extremely variable in the barn owl and its relatives. Variation spans from white to dark reddish and from completely spotless to heavily marked with black feather spots of varying size

A pair of barn owls with the reddish female on the left and the whitish male on the right.
© Alexandre Roulin

ecological niches). A new morph can also spread if it confers benefits when rare. For example, a new morph might enjoy higher foraging success if the prey animals do not recognize it as a predator. Predator–prey relationships can therefore act as a strong selective agent promoting the evolution of colour polymorphisms in raptors and their prey, to improve foraging success or to evade predators.

(2)  **How can colour morphs coexist in the same population?** Colour variation can persist if it makes no difference to the individual whether it displays one morph or another, with no effect on its behaviour or physiology, or on the behaviour of its predators, competitors and potential mates. This possibility is not particularly exciting, and researchers have attempted to determine whether exhibiting a given colouration affects the daily life of its bearer in any way. Scientists wonder whether morphs represent alternative behavioural or physiological strategies with which conspecifics produce the same number of descendants in a lifetime. In other words, they have the same long-term **fitness** but they achieve it by different means.

Two scenarios can be proposed to explain why a morph does not invade a population by pushing other morphs to extinction. First, each morph can be adapted to a **specific habitat** – for instance, with one morph thriving in open habitats and another in closed habitats. The frequency of these two morphs would then be directly related to the prevalence of each habitat. Second, morphs can achieve the same fitness at the so-called **equilibrium frequency** as long as the morph frequency does not depart from this equilibrium. Any departure from the equilibrium frequency would destabilize the system, causing the overly abundant morph to experience reduced fitness and the other, less abundant morph to experience increased fitness. Equilibrium frequency would rapidly return in this scenario. For instance, in a particular population red and white individuals might achieve the same fitness only if 60% (hypothetical value) of them are red and 40% are white. If the red and white morphs become more and less frequent, respectively, perhaps because most immigrants coming from another population are red, red individuals would now experience a reduction in fitness and white individuals an increase in fitness. As a result, the red morph would become less frequent and the white morph more frequent until the 60–40 equilibrium is once more attained. This so-called **frequency-dependent selection** is very powerful for ensuring that one morph does not displace other morphs. For example, to have access to potential mates, black male ruffs are aggressive against competitors, whereas white ruffs adopt sneaky behaviour. If black males become too abundant, they spend too much time competing and less time copulating, which favours sneaky males that focus their attention on attracting females.

(3)     **What is the adaptive function of each morph?** When colour morphs are adapted to specific local environmental conditions, or when fitness is frequency-dependent, morphs can be considered as alternative strategies. Morphs may behave differently to attract potential mates, as do black and white ruffs. Morphs may also represent different reproductive strategies – as in the tawny owl, where reddish males maintain constant brood-raising regardless of environmental conditions, whereas lighter males adjust paternal care in relation to changes in environmental conditions. Differently coloured individuals may also represent alternative physiological strategies, with, for example, one morph physiologically coping with stressful factors better than another morph. This may be the case if morphs differentially regulate the production of glucocorticoids (the stress hormones) in reaction to predators, pathogens, food depletion or other sources of stress.

## FUTURE RESEARCH

- Identifying morph-specific strategies is not easy because these strategies can be context-dependent. Under most conditions, morphs perform similarly, but when it is cold, when food is limited or when there is a pathogen outbreak, one morph might suddenly be more successful. Researchers must therefore measure the reproductive success, survival, behaviour and physiology of differently coloured individuals under various environmental conditions and over time. Once the adaptive function of a morph has been identified, the next step is to understand how different morphs can coexist, i.e. whether a morph is adapted to specific environmental conditions or whether the fitness of each morph varies in relation to its frequency.
- Determining how morphs evolved may require the comparison of multiple species to identify which colouration is ancestral and which is derived.

## FURTHER READING

Literák, I., Roulin, A. and Janda, K. 1999. Close inbreeding and unusual melanin distribution in barn owls (*Tyto alba*). *Folia Zool.* **48**: 227–231.

Roulin, A. 2004. The evolution, maintenance and adaptive function of genetic colour polymorphism in birds. *Biol. Rev.* **79**: 815–848.

These four young barn owls (*on the right*) closely resemble their parent (*on the left*) because of shared genes.

## 12.2 GENETICS OF PLUMAGE POLYMORPHISM

# Heritage

**The number and identity of genes and their impact on the expression of melanin-based colour traits are important for understanding why plumage traits are so variable in the barn owl, grass owl and masked owl. In the barn owl, the strong resemblance between related individuals is due to shared genes and not to shared environment. Because the inheritance of reddish colouration and that of the number and size of black spots are not completely independent, selection exerted on one plumage trait affects the evolutionary trajectory of the two other traits.**

'Quantitative genetics' aims to determine the relative contributions of genetics and the environment to phenotypic variation. For instance, the pink colouration of flamingos depends mostly on the amount of carotenoids obtained from the environment (e.g. the algae and shrimps they eat). In colour-polymorphic animals such as feral pigeons and ruffs, on the other hand, colour differences result from individuals possessing alternative alleles at the polymorphic gene.

**Quantitative genetics** The underside of the barn owl's body varies from white to dark reddish and from spotless to heavily spotted. In order to investigate whether genetics or rearing conditions explains sibling resemblance, cross-fostering experiments were performed in Switzerland by swapping eggs or hatchlings between randomly chosen nests. This allowed us to compare related individuals reared in different environments, and unrelated individuals reared in the same environment. Because differently coloured parents may exploit habitats that differ in quality and may invest differentially in parental care, allocating offspring to randomly chosen nests ensured that nestlings born from parents with different plumages experienced, on average, similar rearing conditions.

The results from these experiments are clear. Full siblings raised in different nests resemble each other with respect to reddish colouration and number and size of black spots, while unrelated nest-mates are not similarly plumaged and resemble their biological parents but not their foster parents. This indicates that related individuals display similar plumage because they share genes and not because they live in the same environment. Moreover, experimentally affecting nestling body condition by increasing brood size did not lead to the production of different plumages compared to nestlings raised in experimentally reduced broods where parents had fewer mouths to feed. Thus, regardless of the quantity and quality of the food an individual consumed, and regardless of its physical condition, it expressed the genetically programmed colour morph.

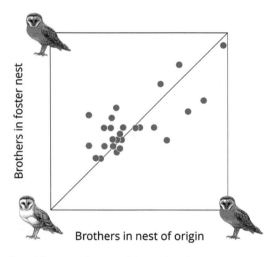

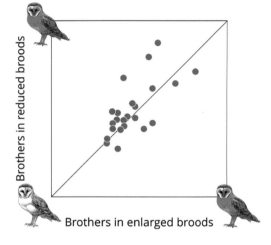

Related barn owls resemble each other with respect to the degree of reddish-brown colouration because they share genes, but not because they share the same rearing environment.
*Above left*: nestling males raised by their biological parents develop similar plumage colouration as their brothers raised by foster parents (dots on the diagonal represent brothers showing exactly the same colouration).
*Above right*: nestling males raised in broods in which two hatchlings were experimentally added developed a similar colouration as their brothers raised in broods in which two hatchlings were removed (dots are as often located above as below the diagonal).
*Right*: nestling females raised in the nest of origin did not resemble their unrelated female nest-mates (dots are not located close to the diagonal, which is the line of equality). Each dot represents the mean sibling reddish colouration.
From Roulin *et al.* 1998.

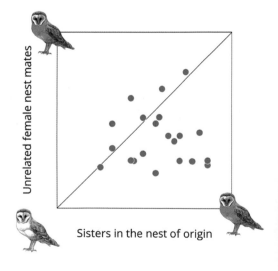

To be differently coloured, it suffices to have a different allele at a polymorphic gene that is responsible for melanin production (i.e. a melanogenic gene). Even if only one or a few genes can modify colouration, in most situations, many genes are involved in the expression of plumage traits. Even without knowing the identity of the particular genes that affect a trait, we can statistically estimate the magnitude of their added effects in differences in plumage traits between individuals with the parameter **heritability**. This parameter ranges from 0 (related individuals do not resemble each other, or genetic factors are not involved in their resemblance) to 1 (perfect resemblance between relatives due to genes). Heritability estimates are larger for reddish colouration ($h^2_{male} = 0.87$; $h^2_{female} = 0.79$) than for spot diameter ($h^2_{male} = 0.81$; $h^2_{female} = 0.58$) and number of spots ($h^2_{male} = 0.72$; $h^2_{female} = 0.50$), indicating that related individuals are more similar with respect to reddish colouration than plumage spottiness.

These heritability estimates are very high, indicating that in the barn owl the expression of plumage traits is under strong genetic control and weakly sensitive to the environment. This means that if individuals displaying a given plumage are in poorer physiological condition or exploit different habitats than differently plumaged conspecifics, this is not because the environment regulates the expression of melanin-based colouration. Rather, a link between plumage, environmental variables and body condition would indicate that displaying a given plumage somehow directly or indirectly influences how individuals exploit their environment or their capacity to withstand stressful environmental events. The expression of a genetically inherited plumage trait could therefore predict their behaviour and physiology. For instance, dark reddish owls and whitish owls might be adapted to different environmental conditions, and might preferentially exploit different prey, in different habitats. As we will see later, this is indeed the case!

The larger heritability estimates for males than for females indicate that the resemblance between parents and offspring is stronger with fathers than with mothers. This is in part because some of the genes encoding plumage traits are located on the sex chromosome Z, with homogametic males having two Z copies and heterogametic females having one Z copy transferred to sons and one copy of the W chromosome transferred to daughters. In birds, the Z chromosome typically contains more genes than the W chromosome, which may therefore have little effect on plumage traits. The finding that genes located on the Z chromosome explain approximately half of the variation in spot size compared to genes located on autosomal chromosomes (27% vs. 44%) is not surprising given that barn owls have 92 chromosomes on which genes encoding for plumage traits could be located. Thus, approximately one-third of the genes (or one-third of their impact) implicated in the production of feather spots are located on the Z sex chromosome and two-thirds are on autosomal chromosomes.

## Genetic correlations between plumage traits

Reddish colouration is naturally selected through predator–prey interactions, and plumage spottiness is sexually selected through mate choice. However, these plumage traits are not independently inherited, but genetically correlated. Thus, natural selection exerted on reddish colouration should also affect the evolution of the sexually selected plumage spottiness, and vice versa. For example, dark reddish owls display more and larger black spots than whitish owls and hence, if natural selection favours reddish owls over whitish owls, not only reddish owls but also heavily spotted owls should increase in frequency. This might be a disadvantage if sexual selection favours small-spotted over large-spotted individuals. Thus, genetic correlations between phenotypic traits are usually considered an evolutionary constraint because genetic correlations indirectly drive organisms beyond their phenotypic optimum.

Determining the exact relationship between plumage traits could provide information on which trait combinations are the most frequent and potentially the best. In the barn owl, reddish colouration and plumage spottiness are based on the deposition of melanin pigments, which is why melanogenic genes could simultaneously alter more than one plumage trait. In line with this statement, the genetic correlation between reddish colouration and the number and size of spots is 0.50 and 0.56, respectively (a value of 0 indicates that the two traits are not genetically correlated, and a value of 1 that the two traits are perfectly correlated). This means that reddish owls have, on average, more and larger black spots

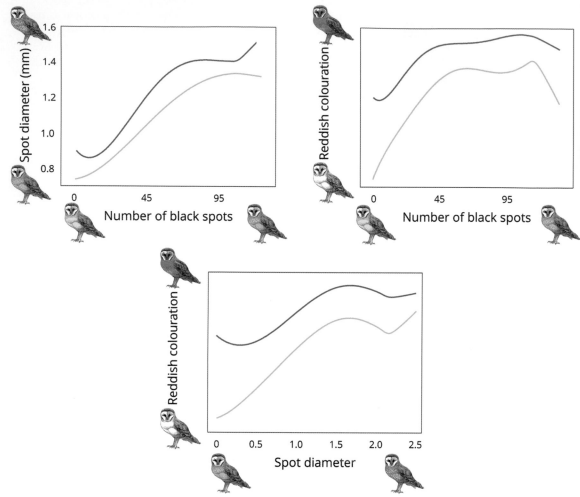

Plumage traits are not independently inherited from each other, as shown by the correlation between plumage traits within 1692 females (*orange line*) and 1547 males (*grey line*) measured in Switzerland. Genetic correlation between the number and size of black spots (*above left*), between the number of black spots and reddish colouration (*above right*) and between the size of black spots and reddish colouration (*below*).

than do white owls. Because the correlation is not perfect (e.g. some white owls can display many large spots), we can conclude that different genes are involved in the development of reddish colouration and plumage spottiness. The genetic correlation between the number and size of black spots is much stronger (0.93), indicating that the more spots a bird inherits, the larger these spots are. The fact that genetic correlations between plumage traits are 1.6 times stronger in males than in females suggests that being either reddish and heavily spotted or white and spotless is more important in males than in females.

## Genetic correlations between sexes
Male and female siblings can resemble each other because they share large parts of the genetic machinery passed on by their parents. They can however differ in important aspects owing to the effects of the sex chromosomes, which mean that for some traits sons more closely resemble their father and daughters their mother. This is not trivial, because in many birds the expression of plumage traits is sex-specific – such as in the ostrich, in which only males are black and females are brown.

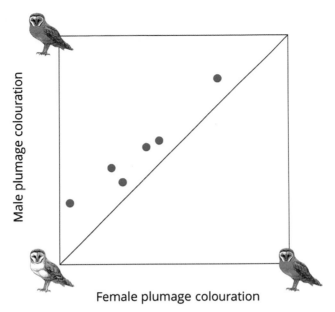

Mean plumage colouration of males and females in six populations of European barn owls. This figure shows that in populations in which females are dark reddish, males are also dark reddish, and in populations in which females are whitish, males are also whitish. From Roulin *et al.* 2001.

We can estimate to what extent a given trait is genetically correlated between the sexes, i.e. to what extent genetic variation is passed on and expressed in both male and female siblings. In the barn owl, the genetic correlations between the sexes for reddish colouration (0.996), spot diameter (0.963) and number of spots (0.903) are close to the maximum value of 1. This means that in the barn owl the same genes are involved in the expression of plumage traits in males and females and that these genes are only slightly differentially expressed in the two sexes. There is thus little variation for genetic factors driving sex-specific effects. For example, if males (but not females) are selected to display a whiter plumage, the evolution of a whitish colouration may proceed slowly because females constantly pass on to their sons genes encoding for a reddish colour; furthermore, females may progressively evolve towards a whitish colouration even if they are not selected to be so, simply because fathers pass on genes encoding for a white plumage to their daughters.

The genetic correlations between the sexes have many implications. Recall the genetic conflict that arises from males and females being positively selected when displaying small and large black spots, respectively. In males, it is better to be spotless whereas in females it is better to be heavily marked with black spots. The strong genetic correlation between the sexes for spot diameter is at the core of that conflict, given that selection exerted on a plumage trait in one sex will inevitably affect the evolution of the same trait in the other sex. This was shown in Switzerland, where positive selection on the size of black spots displayed by females resulted in an increase in the proportion of large-spotted females and males, even if large spots tended to be counter-selected in males.

**Molecular genetics** Traditionally, researchers have used the **candidate gene approach** to identify genes responsible for phenotypic variation. This approach is simple. If one gene is known to participate in the elaboration of melanin pigments in laboratory animals such as mice, researchers working with other organisms, such as the barn owl, will examine whether this very same gene can explain why some owls are dark while others are light. The idea is to investigate whether the gene is polymorphic, with dark birds possessing another allele than white conspecifics. For

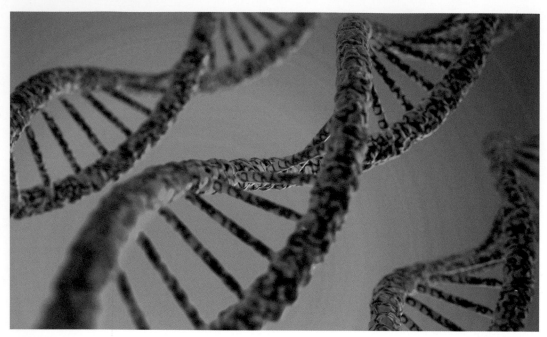

Molecular genetics aims at identifying genes involved in phenotypic traits, such as plumage traits in the barn owl.

instance, this method was used to determine whether the melanocortin-1-receptor ($MC1R$) gene encodes the expression of colour in the barn owl, as it does in many other vertebrates (such as red hair in humans).

In European populations of the western barn owl, $MC1R$ is polymorphic, with the ancestral allele $MC1R_{White}$ encoding a whitish colouration and the derived allele $MC1R_{Reddish}$ encoding a reddish colouration. The later allele evolved during the recolonization of the northeastern parts of the continent after the last Ice Age, and it is still found there at a high frequency. The impact of $MC1R$ on plumage traits varies between body parts and differs between sexes and plumage traits. This gene has a relatively strong effect on the expression of reddish colouration, particularly under the wings of males, where it explains 79% of the variation in colouration, and much less on the breast of females (10%). The impact of $MC1R$ on the number and size of black spots is much less pronounced, explaining less than 1% of the inter-individual variation in spottiness.

The effect of this gene on plumage traits is almost twice as great in males as in females, explaining 64% and 34% of the inter-individual variance in reddish colouration, respectively. This is due to a stronger effect of $MC1R_{White}$ on male colouration than on female colouration, whereas males and females possessing the $MC1R_{Reddish}$ allele are more similar in colour. As a consequence, sexual dimorphism is more pronounced among homozygous individuals possessing two copies of $MC1R_{White}$ than among homozygous individuals possessing two copies of $MC1R_{Reddish}$ and among heterozygous individuals. This pattern of sexual dimorphism is not restricted to Switzerland but is found worldwide, with males and females more closely resembling each other in regions where birds are dark reddish than in those where birds are whitish.

## FUTURE RESEARCH

- Is it better to be dark reddish with many large spots, or to depart from this general pattern (i.e. to be reddish with small spots or white with large spots)? Do black spots improve the potential cryptic function of reddish plumage, or are black spots better seen by conspecifics in white than reddish owls? A number of approaches can be used to evaluate these questions, such as testing whether different combinations of reddish colouration and spottiness are differentially associated with survival and reproductive success. Another approach would be to examine globally whether different ecological factors drive the evolution of different trait combinations.

- Genomics is a new method of screening genomes in a quest to detect genetic variants related to phenotypic variation without any *a priori* knowledge about which gene could be involved in colour production (as is the case with the 'candidate gene' approach). This method may help identify genes implicated in the expression of the different plumage traits beyond *MC1R*.

- *MC1R* should be sequenced across Tytonidae species to determine whether it is implicated in colour variation outside Europe. It would also be interesting to examine whether the same or different mutations are related to colour variation across the range of Tytonidae.

## FURTHER READING

Burri, R., Antoniazza, S., Gaigher, A., Ducrest, A.-L., Simon, C., The European Barn Owl Network, Fumagalli, L., Goudet, J., and Roulin, A. 2016. The genetic basis of color-related local adaptation in a ring-like colonization around the Mediterranean. *Evolution* **70**: 140–153.

Larsen, T. C., Holland, A. M., Jensen, H., Steinsland, I. and Roulin, A. 2014. On estimation and identifiability issues of sex-linked inheritance with a case study of eumelanic spot diameter in Swiss barn owl (*Tyto alba*). *Ecol. Evol.* **4**: 1555–1566.

Roulin, A. 2016. Evolutionary trade-off between naturally and sexually selected melanin-based colour traits in the worldwide distributed barn owls and allies. *Biol. J. Linn. Soc.* **119**: 455–476.

Roulin, A. and Dijkstra, C. 2003. Genetic and environmental components of variation in eumelanin and phaeomelanin sex-traits in the barn owl. *Heredity* **90**: 359–364.

Roulin, A. and Jensen, H. 2015. Sex-linked inheritance, genetic correlations and sexual dimorphism in three melanin-based color traits in the barn owl. *J. Evol. Biol.* **28**: 655–666.

Roulin, A., Richner, H. and Ducrest, A.-L. 1998. Genetic, environmental and condition-dependent effects on female and male plumage ornamentation. *Evolution* **52**: 1451–1460.

Roulin, A., Riols, C., Dijkstra, C. and Ducrest, A-L. 2001. Female- and male-specific signals of quality in the barn owl. *J. Evol. Biol.* **14**: 255–266.

San-Jose, L. M., Ducrest, A.-L., Ducret, V., Béziers, P., Simon, C., Wakamatsu, K. and Roulin, A. 2015. Effect of the *MC1R* gene on sexual dimorphism in melanin-based coloration. *Mol. Ecol.* **24**: 2794–2808.

Males are, on average, whiter than females and display fewer and smaller black feather spots.

## 12.3 SEXUAL DIMORPHISM IN PLUMAGE TRAITS

# Androgyny

**In all barn owl, grass owl and masked owl populations, females are, on average, darker reddish than males and display more and larger black feather spots. However, plumage traits cannot be used to identify the sex of a bird, because some males display typical female plumage and some females display typical male plumage. Sexual dimorphism in colouration is incomplete and therefore varies between families, with brothers sometimes resembling sisters.**

The extent to which males and females are morphologically distinct, or sexually dimorphic, depends on whether different phenotypes are selected in males and females. In birds, females are usually more intensely involved in parental care than males, and for this reason, cryptic colourations are naturally selected in females so that they can remain hidden with their offspring. In contrast, conspicuous colourations are primarily sexually selected in males, to attract a partner or to deter competitors. This common situation may be slightly different in raptors, including owls, in which the male is strongly involved in offspring rearing, implying that male plumage colouration could be naturally selected to ensure high foraging success.

**Sexual dimorphism in plumage traits** All barn owls, grass owls and masked owls are sexually dimorphic with respect to plumage traits. Compared to males, females are, on average, darker reddish and display more and larger black spots on the underside of the body. However, sexual dimorphism is not complete because male-like females (whitish with few small spots) and female-like males (reddish with many large spots) have been observed. In Switzerland, sisters are darker reddish and display more and larger spots than their brothers in only 57% of the nests. Plumage traits can thus not be reliably used to identify sex.

When parents, particularly the mother, are more melanic (i.e. dark reddish and heavily marked with large black spots), their sons and daughters display a similarly melanic plumage. In contrast, when both parents are pale reddish and have lightly spotted plumage, sons are pale and lightly spotted, while daughters are dark and heavily spotted. This means that offspring sexual dimorphism is more pronounced when parents are weakly melanic – and the reason for this is that female offspring do not produce a typical male-like plumage (white and unspotted) rather than because the male offspring do not produce a typical female-like plumage (reddish and heavily spotted). This suggests that it is more costly for females to display a male-specific plumage than for males to display a female-specific plumage. This is a global pattern across barn owl populations, as males and females look more similar where birds are, on average, reddish rather than whitish. For instance, in Germany, where barn owls are dark, sexual dimorphism in plumage traits is much weaker than in Spain, where barn owls are pale.

**Determinants of sexual dimorphism** As shown in Switzerland, the degree of sexual dimorphism in plumage traits is sensitive to genes located on autosomal chromosomes rather than on the sex chromosomes. Sexual dimorphism in reddish colouration is more pronounced than sexual dimorphism in number of spots or their diameter: in 92% of families, sisters are darker reddish than their brothers, with 78% displaying more black spots and 67% displaying larger black spots. This is, however, not a universal pattern. In the United Kingdom, sexual dimorphism in reddish colouration is more pronounced, followed by spot diameter and then spot number, whereas in the USA, the strongest sexually dimorphic traits are spot size and spot number, followed by reddish colouration. This suggests that unknown ecological factors may select for different degrees of sexual dimorphism in different regions.

The only available evidence on this issue indicates that sexual dimorphism in spot size is stronger on remote islands than on islands located close to a continent; sexual dimorphism in the degree of reddish colouration is less pronounced on islands than on the mainland and more pronounced in the southern than the northern hemisphere; sexual dimorphism in the number and size of black spots is more pronounced in temperate zones than near the equator, whereas the opposite pattern is observed for reddish colouration, with females resembling males more closely in temperate zones than near the equator. These results are fascinating – but research is still necessary to understand their biological meaning.

## FUTURE RESEARCH
- The genes involved in the extent to which males and females differ in plumage should be identified and sequenced in many individuals around the world. This would help to determine why, from a mechanistic point of view, sexual dimorphism varies geographically and how it could evolve and be maintained.
- Sexual dimorphism varies between families, raising the question of whether adults rearing daughters that differ in plumage from their brothers achieve a higher fitness than adults rearing similarly coloured sons and daughters. Because sexual dimorphism results from females not producing a white and spotless plumage, rather than from males not producing a dark and heavily spotted plumage, is selection more strongly exerted against male-like females than against female-like males?

## FURTHER READING
Roulin, A. and Jensen, H. 2015. Sex-linked inheritance, genetic correlations and sexual dimorphism in three melanin-based color traits in the barn owl. *J. Evol. Biol.* **28**: 655–666.

Roulin, A. and Randin, C. 2015. Gloger's rule in North American barn owls. *Auk* **132**: 321–332.

Roulin, A. and Randin, C. 2016. Barn owls display larger black feather spots in cooler regions on the British Isles. *Biol. J. Linn. Soc.* **119**: 445–454.

Roulin, A. and Salamin, N. 2010. Insularity and the evolution of melanism, sexual dichromatism and body size in the worldwide-distributed barn owl. *J. Evol. Biol.* **23**: 925–934.

Regardless of sex, barn owls become whiter after the first moult, during the first year of life.

## 12.4 AGE-RELATED CHANGES IN PLUMAGE TRAITS

# Costume change

Although young barn owls already express the full diversity of plumages found in adults, the underside of the body changes colour slightly between the first and second years of life. After the first moult, in both sexes, the newly acquired feathers are whiter than the feathers produced at the nestling stage. In males, the new feathers are less heavily spotted than the feathers they replaced, whereas in females the newly produced black feather spots are larger than the previous spots. Adults achieve greater reproductive success if they become whiter, lose more spots or enlarge their spots compared with the previous moult.

In birds, juveniles frequently exhibit a less conspicuous plumage than that of the adults, which is often but not always acquired at the first moult. Juveniles display a cryptic plumage to avoid predators, or a less sexually attractive plumage to avoid aggression from dominant conspecifics. This situation prevails in many birds – but probably not in the barn owl, in which all plumage types are found in juveniles and adults, even though the plumage progressively changes with age.

**Change in plumage** Barn owls become lighter reddish after the first moult and continue to become lighter with each subsequent moult but to a much lesser extent. While the lightening of the plumage is similar between the two sexes, the change in plumage spottiness is sex-specific: after the first moult, males lose a few black spots, while the spots on females become larger. These changes are relatively minor, and a heavily spotted and reddish fledgling will retain these plumage characteristics into adulthood. Therefore, plumage traits measured in the first year of life provide a good estimate of how plumage will look later in life. Plumage traits are of little help in identifying either the sex or the age of a barn owl.

The change in plumage characteristics is not the result of feather wear, which is much less pronounced in owls than in diurnal raptors, which spend most of the day under damaging ultraviolet sunlight. Thus, when barn owls replace their feathers during their first moult, the new feathers are slightly different not so much because of feather wear but probably mainly because the hormonal profile changes from the juvenile to adult stages. While these unidentified hormones may boost the size of black spots through a higher production of eumelanin, they simultaneously reduce the production of phaeomelanin around black feather spots, so that owls become less reddish with age. Indeed, females whose spots increase in size between two successive moults also become less reddish. The extent of plumage changes clearly has a genetic basis, because a female who becomes whiter than another female between the first and second years of life produces offspring that will also become whiter between these two age classes.

Because hormones regulate morphological traits and sexual activity, not surprisingly, in adulthood, the extent of plumage changes between two years is associated with a concomitant change

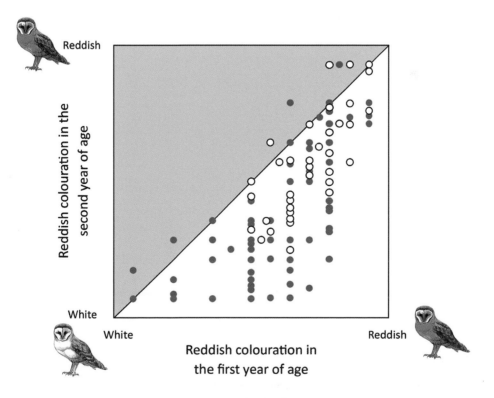

Within-individual change in plumage colouration between the first and second years of life. Each data point represents an individual, with females indicated by orange dots and males by white dots. Points below the diagonal indicate barn owls becoming whiter between these two age classes, and points above the diagonal indicate birds becoming redder.

in reproductive activities. Females whose black spots increase in size between two years reproduce earlier in the season and lay larger eggs; males who lose more black spots between two years raise more offspring; and regardless of sex, individuals who become lighter reddish between two years raise larger families. This suggests that in Switzerland, where these observations were made, whiter barn owls, females displaying larger spots and males displaying fewer spots have selective advantages over other individuals or, alternatively, invest more energy in reproductive activities.

## FUTURE RESEARCH

- The change in plumage with age has been studied only in the barn owl. It should also be studied in other Tytonidae.
- The hormones and genes triggering a change in plumage traits between the first and second years of life should be identified.
- Whether the change in plumage characteristics between the first and second years of life is adaptive should be investigated. We can predict that in both males and females, a dark reddish plumage is beneficial, particularly in the first year of life, whereas the display of few spots in males and of large spots in females is more advantageous in adults than in yearlings.

## FURTHER READING

Dreiss, A. N. and Roulin, A. 2010. Age-related change in melanin-based coloration: females that become more female-like and males more male-like with age perform better in barn owls (*Tyto alba*). *Biol. J. Linn. Soc.* **101**: 689–704.

A heavily spotted
female barn owl.

## 12.5 MATE CHOICE

# Dressed to kill

**Animals take care in choosing a partner with whom to reproduce. Because barn owl males invest substantially more effort in raising their family, they may select genetically superior females, who may in turn select efficient foragers. There is evidence that a male barn owl chooses his mate and adjusts paternal care based on the degree to which the female is marked with black feather spots.**

In animals, males usually invest less effort in parental care than females and therefore have time to copulate with several partners. However, females are typically reluctant to fully satisfy every male's appetite to copulate and are rather picky when choosing a mate. For females, the effort spent in raising a family may indeed be valuable only if the offspring have been sired by high-quality progenitors, who are potentially identified by conspicuous plumage. This situation is typical of most birds, but what about barn owls, in which males invest so much time and energy in feeding their young? Do males and females both select a mate based on plumage traits?

**Non-random pairing with respect to plumage spottiness** Male barn owls invest substantial time and energy in feeding their family. This is worth the effort as long as the off-spring inherit good genes from their mother; the situation is the opposite of what is seen in many animals, in which males do not take care of their offspring and females try to select a mate with good genes. Thus, in the barn owl, we can predict mate choice by males without excluding female mate choice. Because females are more heavily spotted than males, the latter may assess female plumage

Reddish breast feathers with black markings of different sizes and shapes. In reddish individuals, black spots often have a white marking above.

spottiness to choose a high-quality partner, as suggested by the following observations collected in Switzerland:

- Do male and female partners resemble each other (**assortative pairing**) with respect to the size of black spots? While this is the case in some years, in other years large-spotted individuals are paired with small-spotted individuals (**disassortative pairing**). These non-random patterns of pairing do not provide proof that plumage spottiness is a mate choice criterion, as it could also be the result of the non-random distribution of barn owls across habitats. For instance, heavily spotted barn owls may be located in different habitats than lightly spotted owls, leading to assortative pairing.
- Do males repeatedly pair with females displaying black spots of a similar size? The evidence shows that males secure a new partner whose spots are similar to those of the previous mate. This suggests that each male prefers females displaying spots of a specific size. An alternative explanation is that each male defends a nest site that is suitable to females displaying black spots of a specific size, so that year after year each male ends up pairing with similarly looking females.
- Do males adjust their reproductive behaviour according to the plumage spottiness of their mate? To answer this question, female plumage spottiness was experimentally manipulated and the behavioural response of their male mate was monitored. One month after black feather spots of incubating females were cut off with scissors, their partner delivered fewer prey items for the nestlings than did males paired with intact females. This resulted in decreased offspring body mass and, in turn, poorer fledging success. This experiment suggests that males **invest differentially** to rear their offspring in relation to mate spottiness. Although removing black spots did not affect female survival, spotless females bred less often the following year than did spotted ones, suggesting that they experienced difficulties in finding a partner.

## Random pairing with respect to reddish colouration? The underparts of
male barn owls, grass owls and masked owls are less reddish than those of females, and hence white colouration could be considered a masculine plumage trait. Do females assess the degree to which males are whitish when selecting a mate? Answering this question is not easy, because it is difficult to

A barn owl stretching its wing. © Guillaume Rapin

experimentally modify plumage colouration over the entire body underside, i.e. changing the colouration of a reddish owl to white (or vice versa) to examine whether this change alters foraging behaviour or the probability of securing a high-quality mate. The only available evidence comes from France, Germany, Hungary, Israel and Switzerland, where white males pair with white or reddish females as often as reddish males. The absence of assortative pairing with respect to plumage colouration is, however, not proof that this trait is not used as a criterion for mate choice. The mystery remains.

## FUTURE RESEARCH
- Whether owls with different plumage traits secure mates of different qualities should be examined. For example, larger-spotted females could pair more frequently with larger males or with males having higher survival prospects. Because plumage traits vary geographically, the relationship between plumage traits and mate quality may vary between populations.
- The preference for specific plumage traits displayed by the other sex should be formally tested. In addition, it would be informative to investigate whether this changes with age – are yearlings and adults paired with differently plumaged mates?
- Female plumage spottiness was manipulated by cutting off feather tips containing black spots. A similar experiment should also be performed in males.
- Do all males have the same preference for large-spotted females, or does male preference for specific females vary spatially and temporally? More information on mate choice criteria used by male and female barn owls is required.

## FURTHER READING
Roulin, A. 1999. Nonrandom pairing by male barn owls *Tyto alba* with respect to a female plumage trait. *Behav. Ecol.* **10**: 688–695.

Smaller black spots

## 12.6 SEXUALLY ANTAGONISTIC SELECTION

# Transgender barn owls?

**Males and females share most of their genetic machinery and hence closely resemble each other in many aspects. However, a phenotype that is beneficial to one sex may be detrimental to the other sex. In Switzerland, female and male barn owls displaying large black feather spots have higher and lower survival probabilities in the first year of life, respectively. This pattern of positive selection of plumage spottiness in females and negative selection in males is an example of 'sexually antagonistic selection'.**

Although male barn owls are, on average, differently coloured than females, members of both sexes can express any plumage trait. This provides an opportunity to study the impact of resembling the other sex, i.e. being dark reddish and heavily spotted in males and spotless white in females. While the strength of natural and sexual selection is often sex-dependent, this selection can sometimes be reversed, with males and females enjoying high fitness returns when expressing opposite versions of a trait. For instance, in humans, the hip is narrower in males for running faster while hunting and wider in females for giving birth to large-headed children. The problem is that parents pass on their genes to both sons and daughters. As a consequence, when humans were hunter–gatherers, smaller-hipped males produced maladapted daughters, and larger-hipped females produced maladapted sons. Sexually antagonistic selection could therefore account for the maintenance of variation in hip size in men and women.

Larger black spots

Males are selected to be weakly spotted, as indicated by the blue arrow, and females are selected to be heavily spotted, as indicated by the orange arrow.

## Evidence for sexually antagonistic selection

Is it a disadvantage for males to be heavily spotted like females, and for females to be lightly spotted like males? In Switzerland, females displaying large black feather spots have a stronger survival advantage in their first year of life than do smaller-spotted females, while in males the opposite trend prevails, albeit at a smaller magnitude. Stronger positive survival selection in females than negative survival selection in males resulted in an increase in the frequency of large-spotted individuals in the population over a period of twelve years.

From a genetic point of view, sexually antagonistic selection implies that the allele(s) responsible for producing large spots are favoured in females and that the allele(s) encoding small spots are favoured in males. This leads to a **genetic conflict**. The situation is even more complex in the barn owl because the 'small-spotted allele' inherited from the mother, but not from the father, confers a survival advantage in sons, while the 'large-spotted allele' inherited from the father, but not from the mother, confers a survival advantage in daughters. To diminish the intensity of this genetic conflict, barn owls produce more offspring of the sex that benefits the most from inheriting a specific allele. Accordingly, the first-laid egg of small-spotted females is more often a son, and when laid by the mate of large-spotted males, it is more often a daughter.

Is it better to possess the allele encoding large spots or that encoding small spots? Answering this question would improve our understanding of whether the genetic conflict is more intense when parents are large- or small-spotted. In other words, do the benefits derived by large-spotted daughters compensate for the disadvantages accrued by large-spotted sons, and do the benefits derived by small-spotted sons compensate for the disadvantages accrued by small-spotted daughters? The graph

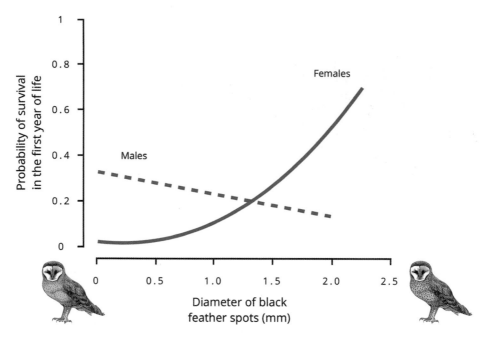

In Switzerland, females displaying large black spots have a higher survival probability in the first year of life than do females with small spots (*orange line*), whereas in males the opposite trend is observed (*broken blue line*).

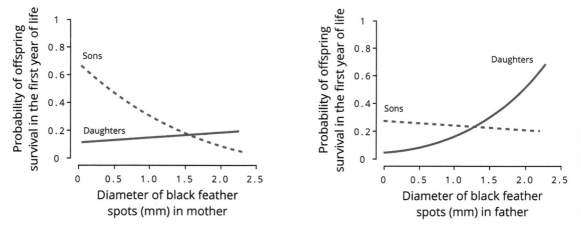

Sons show a higher survival probability in the first year of life if their mother displays small rather than large black spots (*left*), whereas daughters have a higher survival probability if their father displays large rather than small black spots (*right*).

above (top) shows that the difference in survival between males and females is highest when individuals are large-spotted. Genetic conflict may thus be maximal when parents transmit to their progeny the allele encoding large spots. This would be particularly true when the mother is large-spotted, because offspring sexual dimorphism in spot size would be the weakest, with sons and daughters both displaying large spots; in contrast, when the mother is small-spotted, the sons would display small spots, while their sisters would display large spots. This suggests that mutations that increase the size of black spots are associated with more intense genetic conflict, whereas mutations that suppress the expression of large spots may reduce the intensity of genetic conflict. Sexually antagonistic selection is a rather complex issue!

## FUTURE RESEARCH

- Reddish and whitish plumages are naturally selected, but it remains unknown whether they are also sexually selected.
- The genes experiencing sexually antagonistic selection should be identified to understand why females show a higher survival probability in their first year of life when their father is large-spotted, and why males survive better when their mother is small-spotted.
- Sexually antagonistic selection could explain why we still observe small- and large-spotted individuals. Although large-spotted males could be counter-selected, they may persist because their large-spotted mothers are positively selected and pass on their genes to their sons. Similarly, although small-spotted females are counter-selected, they may persist because their fathers continue to produce small-spotted daughters.
- The intensity of sexually antagonistic selection should be compared between populations in which owls display large spots and those in which they display small spots.

## FURTHER READING

Roulin, A. and Jensen, H. 2015. Sex-linked inheritance, genetic correlations and sexual dimorphism in three melanin-based color traits in the barn owl. *J. Evol. Biol.* **28**: 655–666.

Roulin, A., Altwegg, R., Jensen, H., Steinsland, I. and Schaub, M. 2010. Sex-dependent selection on an autosomal melanic female ornament promotes the evolution of sex ratio bias. *Ecol. Lett.* **13**: 616–626.

Extreme colour variants in the barn owl: white and reddish individuals.

## 12.7 ADAPTIVE FUNCTIONS OF WHITISH AND REDDISH COLOURATION

# With flying colours

**Colouration of the underparts is the trait that varies the most between individuals. In the same population, barn owls can be white or dark reddish or display any colouration between these two extremes. Colour seems to play important roles in the daily life of the barn owl, because it is related to diet, habitat exploitation and reproductive behaviour. Redder owls also disperse longer distances and are more social than white conspecifics.**

Being white or reddish makes a difference. Indeed, colour morphs are adapted to local environmental conditions and play a role in predator–prey interactions, which may have promoted the evolution of colour-specific behaviours.

**Foraging behaviour** Being dark or pale reddish strongly alters the barn owl's appearance and should affect foraging. This is probably why, in Israel and Switzerland, reddish owls eat more voles and whitish ones more mice (murids). In addition, in Israel, white barn owls capture large gerbils more often than reddish owls. To hunt such agile animals, which run very fast, short wings may also help, by making it easier to turn sharply in flight – and indeed, short-winged owls feed more on gerbils than do long-winged individuals. Being both whitish and short-winged should therefore facilitate hunting these animals, and accordingly, lighter-coloured owls have shorter wings in Israel, a property not observed in Europe, where gerbils are absent.

Previous studies have also found that reddish owls fly more often with a full stomach than do white owls, based on the observation that redder individuals had heavier stomach contents when found dead along roads. A heavy body mass while flying may increase pressure on the wings, which must be transmitted from the feathers to the muscles and bones, potentially explaining why the secondary wing feathers of redder birds are anchored deeper inside the integument.

**Habitat choice** Barn owls are predominantly whitish in southern Europe and reddish in northeastern Europe. This geographic variation in colouration persists through the action of natural selection apparently because reddish colouration is adapted to capture voles, which are abundant in northeastern Europe, and whitish colouration is adapted to capture mice, which are more abundant in southern Europe. Because voles live in open habitats and wood mice in wooded areas, reddish barn owls may preferentially breed in open habitats and whitish owls in wooded areas. This is indeed what was observed for females in Israel and Switzerland. This non-random habitat distribution with respect to plumage colouration is adaptive – reddish and whitish females produce more high-quality fledglings when breeding in open and wooded habitats, respectively.

**Reproduction** The long-term coexistence of whitish and reddish owls implies that differently coloured individuals achieve the same fitness in the long term. To date, no study has investigated whether whitish and reddish owls perform differently across a lifetime. We only know that in Switzerland and Israel, the annual reproductive success of whitish and reddish individuals was similar across a few years, suggesting that colour morphs are equally successful. However, in the present context of global environmental changes, including climate change, studies should investigate whether the relationship in performance between colour morphs is changing, which may induce a change in morph frequencies. This is plausible, because in Switzerland white females reproduce on average earlier in the season than reddish females, and in some years reddish males feed their brood at a higher rate than white males. These observations indicate that colouration is associated with reproductive

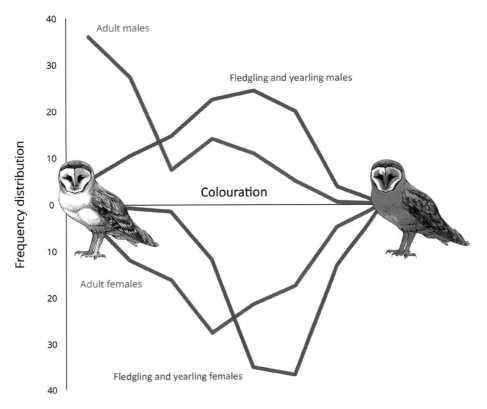

Frequency distribution of different plumage colourations in the barn owl in Switzerland. This figure is based on the measurement of 1987 male fledglings and 2038 female fledglings, 415 male yearlings and 505 female yearlings, and 288 male adults and 327 female adults. It shows that a large proportion of adult males are whitish, whereas younger males and most females are intermediately coloured.

behaviour, implying that differently coloured birds probably adopt alternative reproductive strategies and may therefore be differentially sensitive to variation in environmental conditions.

## Dispersal
The capacity to explore new territories is important for finding a suitable nesting site. Some individuals may disperse longer distances than others because they are less competitive in securing nest cavities close to their natal site or because they take more risks to find better places to reproduce. Two independent studies performed in Switzerland showed that reddish barn owls have an inherent tendency to disperse longer distances than white conspecifics (on average 22 vs. 6 kilometres). Why colouration is related to dispersal behaviour is still unknown, but interestingly, a higher propensity to disperse in darker reddish individuals has also been observed in the barn swallow.

## Social behaviour
The finding that in some years redder barn owls feed their family more often than whitish owls and hence potentially invest more effort in taking care of their young raises the hypothesis that the degree of reddish colouration could be related to social interactions within the family. The data at hand are consistent with this idea. Darker reddish nestlings are more willing to share food with their siblings than are paler nestlings, reddish female nestlings preen their siblings for longer periods of time, and reddish mothers preen their offspring more often than whitish mothers. Therefore, reddish barn owls tend to be altruistic, and whitish owls selfish!

## FUTURE RESEARCH
- Tytonidae vary strongly in colouration worldwide. Which ecological factor(s) promoted the evolution of reddish or white plumage remains to be determined. Such analyses are important, for example to explain why barn owls are whiter in southern Europe and South America and redder in northern Europe and South America. New analyses show that in western, American and eastern barn owls plumage colour becomes darker reddish with increasing annual rainfall, and in American and western barn owls plumage also becomes darker reddish with decreasing temperature. It is still unknown whether the climatic variables causally affect the evolution of plumage colouration in the barn owl, or whether other correlated factors are at play.
- Why do differently coloured birds coexist in the same region? Is it because colour morphs are adapted to different habitats, or because individuals of a counter-selected morph regularly immigrate from other regions where they thrive?
- White and reddish barn owls have adopted different reproductive strategies in Switzerland, with white females reproducing earlier in the season and reddish males sometimes devoting more effort to rearing their offspring. Are these results specific to the 1000-square-kilometre study area in Switzerland or to the study period? More importantly, do white and reddish owls differ similarly in behaviour in other Tytonidae populations?
- Climate change has been shown to differentially affect different coloured morphs in several bird species, such as the tawny owl. Has the performance of reddish and white barn owls changed in recent years due to global warming?
- The reasons why redder barn owls disperse longer distances than whitish owls, and why they are more social, should be investigated.

## FURTHER READING
Antoniazza, S., Ricardo, K., Neuenschwander, S., Burri, R., Gaigher, A., Roulin, A. and Goudet, J. 2014. Natural selection in a post-glacial range expansion: the case of the colour cline in the European barn owl. *Mol. Ecol.* **23**: 5508–5523.
Charter, M., Peleg, O., Leshem, Y. and Roulin, A. 2012. Similar patterns of local barn owl adaptation in the Middle East and Europe with respect to melanic coloration. *Biol. J. Linn. Soc.* **106**: 447–454.

Charter, M., Leshem, Y., Izhaki, I. and Roulin, A. 2015. Pheomelanin-based colouration is correlated with indices of flying strategies in the barn owl. *J. Ornithol.* **156**: 309–312.

Dreiss, A. N., Antoniazza, S., Burri, R., Fumagalli, L., Sonnay, C., Frey, C., Goudet, J. and Roulin, A. 2012. Local adaptation and matching habitat choice in female barn owls with respect to melanic coloration. *J. Evol. Biol.* **25**: 103–114.

Romano, A., Séchaud, R., Hirzel, A. H. and Roulin, A. 2019. Climate-driven convergent evolution of plumage colour in a cosmopolitan bird. *Global Ecol. Biogeogr.* **28**: 496–507.

Roulin, A. 2013. Ring recoveries of dead birds confirm that darker pheomelanic barn owls disperse longer distances. *J. Ornithol.* **154**: 871–874.

Roulin, A., Riols, C., Dijkstra, C. and Ducrest, A.-L. 2001. Female- and male-specific signals of quality in the barn owl. *J. Evol. Biol.* **14**: 255–266.

Roulin, A., Da Silva, A. and Ruppli, C. A. 2012. Dominant nestlings displaying female-like melanin coloration behave altruistically in the barn owl. *Anim. Behav.* **84**: 1229–1236.

Roulin, A., des Monstiers, B., Ifrid, F., Da Silva, A., Genzoni, E. and Dreiss, A. N. 2016. Reciprocal preening and food sharing in colour polymorphic nestling barn owls. *J. Evol. Biol.* **29**: 380–394.

Van den Brink, V., Dolivo, V., Falourd, X., Dreiss, A. N. and Roulin, A. 2012. Melanic color-dependent anti-predator behavior strategies in barn owl nestlings. *Behav. Ecol.* **23**: 473–480.

The black feather spots on the underparts of a barn owl vary in size.

## 12.8 ADAPTIVE FUNCTIONS OF SMALL AND LARGE BLACK SPOTS

# The larger the better

How can reliable information about the intrinsic quality of potential mates or of competitors be obtained? Physiological traits or how much an individual intends to invest in future reproduction cannot be easily assessed without the display of specific behaviour or visible traits. In the barn owl, black spots located on the tips of the feathers of the underparts signal the quality of physiological processes, behaviour and reproductive decisions. In Switzerland, larger-spotted females have a selective advantage over smaller-spotted females, producing offspring that are more resistant to numerous stressful factors, such as parasites and food depletion. This leads to higher female survival in the first year of life.

Given that we observe such large variations in the size of black spots on the underparts of barn owls, do heavily and lightly spotted individuals differ in other respects? The results of numerous studies have shown astonishing connections between the size or number of black feather spots and physiology, behaviour and personality. The very first results that were obtained were unbelievable (to the point that some important scientific journals refused to publish them), but they have now been confirmed so many times that nobody can ignore them any more.

The most frequently used method to study the potential link between plumage spottiness and physiological traits is the cross-fostering experiment. This approach, as well as other methods, has demonstrated that plumage spottiness is associated with aspects of individual quality. Eggs or hatchlings were swapped between nests to ensure that nestlings were reared by randomly chosen foster parents. When the nestlings were close to taking their first flight, several phenotypic traits were measured, such as body size and resistance to stressful environmental factors, measures that were then compared with plumage spottiness in the nestlings themselves and in their foster and biological parents.

Phenotypic traits measured in the cross-fostered nestlings were usually associated only with the size of black feather spots on their biological mother, sometimes with spot size measured in the nestlings themselves, rarely in their biological father, and never in the foster parents. Thus, if spot size signals aspects of individual quality, this is probably because spot size in the parents is associated with genes, or substances packed in the eggs, that they pass on to their offspring, which affect the expression of other phenotypic traits. Unless stated otherwise, all studies described below have been undertaken in Switzerland.

## Stressful factors
Many morphological, behavioural and physiological phenotypes helpful in resisting stress are related to spot size. A young barn owl is commonly more stress-tolerant when it or its biological mother is large-spotted.

- **Resisting stress.** When facing stressful events, such as lack of food, pathogens, predator attacks or oxidative stress, animals quickly release more glucocorticoids in the bloodstream, such as corticosterone in birds and cortisol in humans. This allows individuals to reallocate energy for emergency physiological and behavioural functions to escape the risky situation. While such a response is beneficial in the short term, maintaining high corticosterone levels in the blood is physiologically detrimental. Therefore, stress-induced levels of corticosterone must be finely

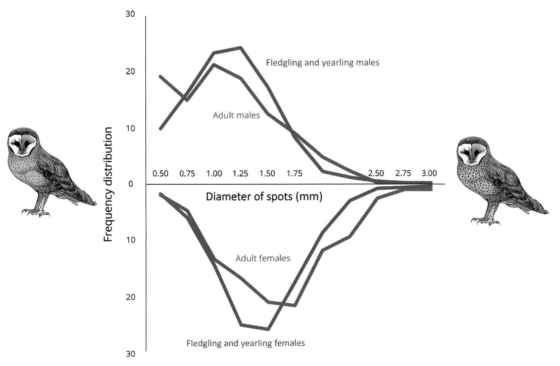

Frequency distribution of the size of black spots on the underparts of barn owls in Switzerland. Most barn owls display spots of an intermediate size, and females have slightly larger spots than males.

adjusted, and as soon as the risky situation has vanished, individuals should experience a return to baseline corticosterone levels.

Nestlings born from larger-spotted mothers mount a less intense corticosterone stress-induced response, and they more quickly reduce the experimentally induced elevation of blood-circulating corticosterone than nestlings born from smaller-spotted mothers. Nestlings born from larger-spotted mothers therefore suffer less from a stress-induced corticosterone response, as measured by body mass and wing growth. Similar results were obtained in female adults, in whom capture induces a weaker rise in corticosterone levels in large-spotted than in small-spotted females.

- **Parasites.** The size of black spots signals the capacity to resist parasites. Compared to offspring of smaller-spotted mothers, those of larger-spotted mothers produce more antibodies specifically directed against a vaccine, host fewer ectoparasitic flies (*Carnus hemapterus*) on their body and have fewer eggs of these insects laid in their nests.
- **Resistance to oxidative stress.** The size of black feather spots signals the ability to resist oxidative stress, the by-product of metabolism that causes damage to all components of the cell. Nestlings displaying larger black spots produce more antioxidants to detoxify the body than smaller-spotted nestlings.
- **Developmental homeostasis.** Stressful rearing conditions can disturb the biosynthesis of the two sides of the body that are under the control of the same genes. For instance, in humans, the left and right sides of the face are often not perfectly symmetrical. The capacity to produce a symmetrical body can therefore reflect the possession of genes that buffer environmental stressful factors. In the barn owl, the size of black feather spots is associated with genes that play such a buffering role. The offspring born from larger-spotted mothers produce a more symmetrical body, with primary and secondary feathers of the left and right wings being more often of the same length than wing feathers of the offspring born from smaller-spotted mothers.
- **Laterality.** Being left-handed or right-handed can be beneficial in some circumstances, and the speed and accuracy of performing a task can be higher in individuals who use their preferred side. However, being lateralized is not always advantageous and can be induced by stress. Although barn owls should preen and scratch the left and right sides of their body to the same extent, nestlings born from small-spotted mothers clean one side more often, whereas nestlings born from large-spotted mothers take care of both sides of the body more equally. This suggests that the capacity of large-spotted owls to withstand stress is related to the degree of behavioural laterality.
- **Anti-predator behaviour.** In the presence of predators, nestling barn owls can scare or attack them, or they can feign death to escape attack. It turns out that the barn owls' reaction to humans is associated with plumage spottiness both in Israel and in Switzerland. When handled, small-spotted individuals are aggressive and agitated, and large-spotted individuals avoid direct confrontation by staying calm, feigning death and trying to scare predators by hissing loudly. Small-spotted owls therefore adopt bold and proactive agitated behaviour, and large-spotted owls are docile.
- **Food deprivation.** Lack of food is a major cause of mortality in nestling and adult barn owls. Selection to resist food deprivation should therefore be strong. When food is provided *ad libitum*, nestlings displaying larger black spots show a lower appetite than smaller-spotted nestlings, and after being deprived of food for 24 hours, nestlings with large black spots lose less body weight. Therefore, the size of black spots reflects the ability to withstand periods of food depletion, given the lower appetite of large-spotted individuals and their ability to cope with lack of food. The capacity to deal with periods of food restriction is also linked to maternal but not paternal spot diameter. This is consistent with the finding that after having eaten at night, the body mass of breeding females in the morning is not correlated with the size of their black feather spots, whereas in the evening, having not eaten during the whole day, larger-spotted females are heavier than smaller-spotted females.

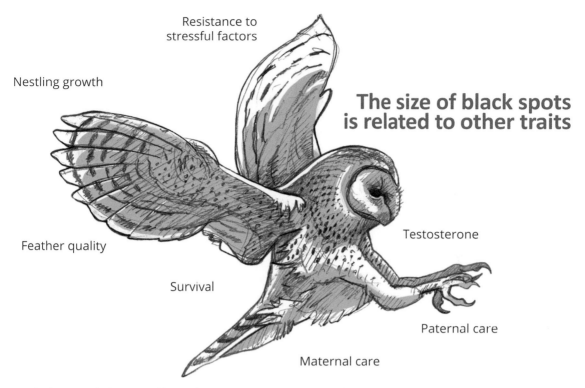

Resistance to
stressful factors

Nestling growth

**The size of black spots
is related to other traits**

Feather quality

Testosterone

Survival

Paternal care

Maternal care

In the barn owl, the size of black feather spots displayed on the underparts covaries with a number of other phenotypic traits, such as the ability to resist stressful factors, nestling growth, feather quality, survival, parental care and testosterone levels.

**Nestling growth** During the resting phase from early morning until the first parental feeding visit of the night, nestlings with large black spots are heavier than those with small spots. This concurs with the above-described observations showing that the size of black spots is related to the capacity to withstand stressful environmental factors. What about the relationship between nestling body mass and the size of black spots measured in their biological parents? Cross-fostering experiments show that in poor years, offspring of smaller-spotted mothers are heavier throughout the daylight hours when they rest and fast. This indicates that small spots displayed by females could be related to genetic factors that are beneficial under certain environmental conditions.

**Testosterone** This hormone mediates many physiological processes and behaviours related to reproductive activities, such as aggressiveness and sexual activity. Although testosterone can trigger the production of melanin, dark individuals may not necessarily have more circulating testosterone than pale individuals. However, the degree of melanin-based colouration of a bird can modify social interactions and, in turn, testosterone levels. Furthermore, genes involved in melanogenesis, such as the pro-opiomelanocortin (*POMC*) gene, also regulate the production of testosterone in interactions with prevailing environmental conditions. Therefore, the link between colouration and testosterone levels could vary through time and space in relation to many factors, including reproductive stage, age, sex and environmental conditions.

In male barn owls, nestlings and breeding adults displaying smaller black spots show higher testosterone levels, which is consistent with the observation that smaller-spotted individuals are more aggressive than larger-spotted conspecifics. In sharp contrast, larger-spotted breeding females have more testosterone than small-spotted females during the rearing period (but not during the incubation period), which is concordant with the fact that larger-spotted females are, on average, more likely to start breeding at the age of one year than smaller-spotted females.

**Maternal care** The size of black feather spots is related to reproductive behaviour. In Israel and Switzerland, larger-spotted females breed earlier in the season and lay larger eggs than smaller-spotted females, and females in whom spot size increases from one year to the next advance breeding date and lay larger eggs the following year. When environmental conditions become difficult for taking care of a family, long-lived animals, such as the barn owl, should reduce reproductive investment to survive until the next breeding season. Even if larger-spotted females invest more effort in reproduction, they are not suicidal! When stressed, large-spotted females shift resources initially devoted to their offspring to self-maintenance as an emergency mechanism. Under prime conditions, large-spotted females brood their offspring more than small-spotted females, but when administered extra corticosterone, a hormone that rises after stressful episodes, large-spotted females reduce brooding investment and delay helping their mate hunt for prey items for the brood.

**Paternal care** In one study, large-spotted males were less efficient foragers, as they travelled longer distances to capture their prey than small-spotted males. Smaller-spotted males thus fed their brood at a higher rate than larger-spotted males. This is consistent with the hypothesis that being small-spotted is favoured in males.

**Feather quality** One of the primary functions of melanin is to protect the integument against physicochemical assault from the environment, such as dust, rain, ultraviolet light and feather-eating parasites. Highly melanized feathers are indeed harder and wear less rapidly than melanin-free feathers. As predicted, if larger-spotted birds are more melanized on other body parts than smaller-spotted conspecifics, their wing and tail feathers are stronger: the base of the shaft (or calamus) of their wing feathers and the vane of their wing and tail feathers are heavier, and their flight feathers contain more melanin and have more transverse black bars, which may strengthen feathers and prevent them from bending too much while flying. Perhaps because their feathers are stronger, heavily spotted barn owls spend less time preening their body feathers. Furthermore, their uropygial gland, which produces wax to clean and protect feathers, is less heavy, suggesting that less wax is required in heavily spotted birds than in lightly spotted birds.

**Survival** A long-term study in Switzerland in which nestlings and breeding adults were ringed and recaptured yearly over twelve years showed that larger-spotted females have higher survival in the first year of life than smaller-spotted females. Such selective advantage may also prevail in other populations, given that in both Israel and Switzerland, larger-spotted breeding females are heavier than smaller-spotted females.

**Honesty of signalling quality** How can an individual barn owl identify that a large-spotted conspecific is more resistant to stress than a small-spotted individual, or that it will produce more stress-resistant offspring? An individual can be certain because the size of black feather spots is genetically inherited and reflects the possession of genes that allow differently spotted individuals to perform differently. Thus, when the environmental conditions are stressful, a male barn owl should preferentially mate with a heavily spotted female to produce resistant offspring, whereas when environmental conditions are less stressful, mating with a heavily spotted female may be less of a necessity.

The genetic link between spottiness and intrinsic quality is so strong that owls cannot act against what their spots indicate. However, the reason for this effect is unknown. Why is this genetic link so strong? One possibility is that melanin pigments have evolved to help organisms resist the damaging impact of ultraviolet light, to fight parasites and pathogens (in insects, melanin encapsulates internal parasites), or to absorb solar energy, allowing animals to warm up more rapidly in cold environments. Resisting stress commonly requires a battery of adaptations; thus, natural selection may have favoured the evolution of melanin and other behavioural and physiological mechanisms. Selection to display all these traits under stressful conditions may be so strong that at some point during evolution these traits (melanin, behaviour and physiology) became genetically correlated, ensuring that these properties were concomitantly inherited.

## FUTURE RESEARCH

- Females displaying larger black spots produce offspring that are more resistant to numerous stressful factors, and which show higher survival prospects in the first year of life. As shown in Switzerland, natural selection favoured larger-spotted females, which resulted in an increase in the frequency of large-spotted individuals in the population over a period of only twelve years. Why do we still find small-spotted individuals in the population? We do not yet have a definite answer. Three non-mutually exclusive mechanisms can be proposed. First, if selection favours large-spotted females in Switzerland, small-spotted females may have an advantage in other populations, and immigration from these populations could replenish Switzerland with small-spotted individuals. Another possibility is sexually antagonistic selection, with large-spotted males being counter-selected and small-spotted males constantly producing small-spotted daughters. Third, the strong positive selection on heavily spotted females may be only transitory, and during other periods, small-spotted females may have a selective advantage so that, in the long term, there is no net benefit to being large-spotted. This might be the case if under some environmental conditions (which did not prevail during the studies performed in Switzerland), large spots are associated with behaviours and physiological processes that entail more costs than benefits. Identifying which of these three mechanisms is responsible for the coexistence of small-spotted and large-spotted females is a challenge for future studies.
- From a mechanistic point of view, why is the size of maternal black feather spots associated with so many behavioural and physiological phenotypic traits, as measured in biological offspring raised in a foster family? Given that the production of small or large spots depends on genetic factors and not on whether individuals are in prime or poor condition, the biological mother may pass on to her offspring genes, epigenetic factors or nutrients in the eggs that determine not only the size of their spots but also other phenotypic attributes. For instance, genes involved in the production of black feather spots may have pleiotropic effects, with one gene regulating several functions. Identifying the genetic mechanism underlying an association between plumage spottiness and other phenotypes is one of the most important undertakings to understand the adaptive function of this plumage trait.

## FURTHER READING

Almasi, B. and Roulin, A. 2015. Signalling value of maternal and paternal melanism in the barn owl: implication for the resolution of the lek paradox. *Biol. J. Linn. Soc.* **115**: 376–390.

Almasi, B., Roulin, A., Jenni-Eiermann, S. and Jenni, L. 2008. Parental investment and its sensitivity to corticosterone is linked to melanin-based coloration in barn owls. *Horm. Behav.* **54**: 217–223.

Almasi, B., Jenni, L., Jenni-Eiermann, S. and Roulin, A. 2010. Regulation of stress-response is heritable and functionally linked to melanin-based coloration. *J. Evol. Biol.* **23**: 987–996.

Almasi, B., Roulin, A., Korner-Nievergelt, F., Jenni-Eiermann, S. and Jenni, L. 2012. Coloration signals the ability to cope with elevated stress hormones: effects of corticosterone on growth of barn owls is associated with melanism. *J. Evol. Biol.* **25**: 1189–1199.

Almasi, B., Roulin, A. and Jenni, L. 2013. Corticosterone shifts reproductive behaviour towards self-maintenance in the barn owl and is linked to melanin-based coloration in females. *Horm. Behav.* **64**: 161–171.

Dreiss, A., Henry, I., Ruppli, C., Almasi, B. and Roulin, A. 2010. Darker eumelanic barn owls better withstand food depletion through resistance to food deprivation and low appetite. *Oecologia* **164**: 65–71.

Dreiss, A. N. and Roulin, A. 2010. Age-related change in melanin-based coloration: females that become more female-like and males more male-like with age perform better in barn owls (*Tyto alba*). *Biol. J. Linn. Soc.* **101**: 689–704.

Roulin, A. 2007. Melanin pigmentation negatively correlates with plumage preening effort in barn owls. *Funct. Ecol.* **21**: 264–271.

Roulin, A., Jungi, T. W., Pfister, H. and Dijkstra, C. 2000. Female barn owls (*Tyto alba*) advertise good genes. *Proc. R. Soc. Lond. B* **267**: 937–941.

Roulin, A., Riols, C., Dijkstra, C. and Ducrest, A.-L. 2001. Female plumage spottiness and parasite resistance in the barn owl (*Tyto alba*). *Behav. Ecol.* **12**: 103–110.

Roulin, A., Ducrest, A.-L., Balloux, F., Dijkstra, C. and Riols, C. 2003. A female melanin-ornament signals offspring fluctuating asymmetry in the barn owl. *Proc. R. Soc. Lond. B* **270**: 167–171.

Roulin, A., Altwegg, R., Jensen, H., Steinsland, I. and Schaub, M. 2010. Sex-dependent selection on an autosomal melanic female ornament promotes the evolution of sex ratio bias. *Ecol. Lett.* **13**: 616–626.

Roulin, A., Mangel, J., Wakamatsu, K. and Bachmann, T. 2013. Sexually dimorphic melanin-based color polymorphism, feather melanin content and wing feather structure in the barn owl (*Tyto alba*). *Biol. J. Linn. Soc.* **109**: 562–573.

San-Jose, L. M. and Roulin, A. 2018. Towards understanding the repeated occurrence of associations between melanin-based coloration and multiple phenotypes. *Am. Nat.* **192**: 111–130.

Stier, K. S., Almasi, B., Gasparini, J., Piault, R., Roulin, A. and Jenni, L. 2009. Effects of corticosterone on innate and humoral immune function and oxidative stress in barn owl nestlings. *J. Exp. Biol.* **212**: 2085–2091.

Van den Brink, V., Dolivo, V., Falourd, X., Dreiss, A. N. and Roulin, A. 2012. Melanic color-dependent anti-predator behavior strategies in barn owl nestlings. *Behav. Ecol.* **23**: 473–480.

The black feather spots on a barn owl's underparts can vary greatly in number, as well as in size.

## 12.9 ADAPTIVE FUNCTIONS OF FEW AND MANY BLACK SPOTS

# To be spotted

In the barn owl, the feathers of the underparts are not all marked with black spots, and some feathers have more than one spot. The pronounced inter-individual variation in spot number is related to sleep and the physiological and behavioural capacity to warm up the body. When ambient temperatures decrease, heavily spotted nestlings increase their resting metabolic rate and huddle more often against their siblings to warm up their own body, which cools down more rapidly than in weakly spotted individuals.

The number of spots that barn owls display on the breast, belly, flanks and underside of the wings varies between individuals: some owls are immaculate, and others display hundreds of spots. Although this pronounced variation is related to fewer phenotypic traits than is spot diameter, weakly and highly spotted individuals adopt different physiological and behavioural strategies to grow from a small chick to a full-grown fledgling.

**Body temperature** Nestling birds have more difficulty than adults in maintaining body temperature. In the barn owl, when the ambient temperature decreases from 30 °C to 15 °C, nestling body temperature drops from 40 °C to 36 °C. Keeping the body warm requires not only food to fuel metabolism but also the capacity to huddle against siblings for warmth. These physiological and behavioural properties are related to the number of black spots displayed on the

The plumage of barn owls varies geographically. For example, in North America they display, on average, more black spots in the centre of the continent than on the west and east coasts.

underparts. Nestlings with more spots have more difficulty warming up, and their bodies are cooler than those of nestlings with fewer spots. To overcome this problem, heavily spotted nestlings have a higher resting metabolic rate and more promptly huddle against their siblings when ambient temperatures become low. These observations raise the fascinating hypothesis that heavily spotted nestlings can reduce the amount of energy used to warm up by adopting a specific behaviour (huddling against siblings). Could the released energy be used to speed up growth and the maturation of organs?

## Sleep and vigilance

Nestlings with more spots have greater difficulty warming up and have a higher resting metabolic rate, and hence they may not have the luxury of missing a parental feeding event. This may be why they need to be highly vigilant in quickly detecting an arriving parent and intercepting the delivered food. Accordingly, nestlings whose mothers display more black spots have shorter bouts of non-REM sleep and more episodes of wakefulness events, and are more vigilant in looking towards the nest entrance where their parents bring food and towards their siblings, against whom they compete. This suggests that to fuel their needy metabolism, spotted nestlings enhance the likelihood of monopolizing parental food resources by being in deep sleep less often and being highly vigilant for when parents bring food to the nest.

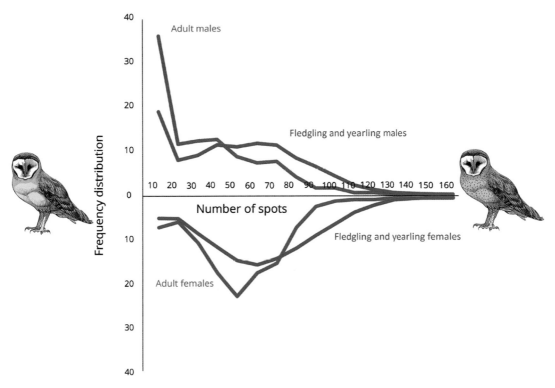

Frequency distribution of the number of black feather spots on the underparts of barn owls in Switzerland. Many males display few spots, whereas most females display, on average, an intermediate number of spots.

## FUTURE RESEARCH

- The genetic mechanism underlying the link between number of spots, thermoregulation and sleep architecture is unknown and should be investigated. Additionally, from an eco-physiological point of view, data are required to explain why heavily spotted nestlings have more difficulty warming up their body than weakly spotted nestlings.
- The relationship between the number of black feather spots and sibling competition should be studied in detail.
- The role of spot number in adults should also be elucidated.

## FURTHER READING

Dreiss, A. N., Séchaud, R., Béziers, P., Villain, N., Genoud, M., Almasi, B., Jenni, L. and Roulin, A. 2016. Social huddling and physiological thermoregulation are related to melanism in the nocturnal barn owl. *Oecologia* **180**: 371–381.

Scriba, M., Rattenborg, N., Dreiss, A. N., Vyssotski, A. L. and Roulin, A. 2014. Sleep and vigilance linked to melanism in wild barn owls. *J. Evol. Biol.* **27**: 2057–2068.

A beautiful reddish
barn owl with many
large black spots.

## 12.10  GEOGRAPHIC VARIATION IN PLUMAGE TRAITS

# Endless forms most beautiful

**Melanism is implicated in thermoregulation and protects the integument against water, ultraviolet light and parasites. For all those reasons, animals located at high and low latitudes are not similarly coloured. This is also the case in the barn owl: in the Americas, Europe, Africa and Australasia, reddish colourations evolved in cold and rainy regions and whitish colourations in warm and dry regions. Geographic variation in the extent to which barn owls are spotted is, in contrast, specific to each continent.**

The cosmopolitan distribution of the Tytonidae offers the unique possibility of testing whether the same ecological and climatic factors repeatedly promoted the evolution of dark versus white plumages or of spotless versus heavily spotted plumages on all continents. To this end, 11 000 Tytonidae specimens preserved in natural history museums, including 8500 barn owls, were measured and analysed.

**Latitudinal variation in plumage traits**  Geographic variation in plumage traits can be pronounced. In the northern parts of the European and North American continents, barn owls are darker reddish and display larger spots than those in the southern parts, although this so-called **cline variation**

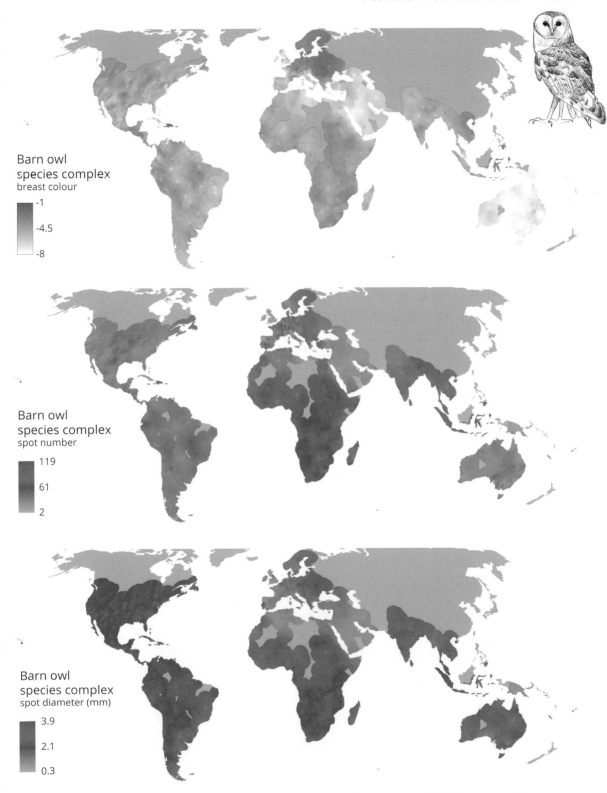

Geographic variation in the degree of reddish colouration, the number of black spots and the size of black spots in the barn owl species complex. For colouration the scale is from white (–8) to dark reddish (–1).

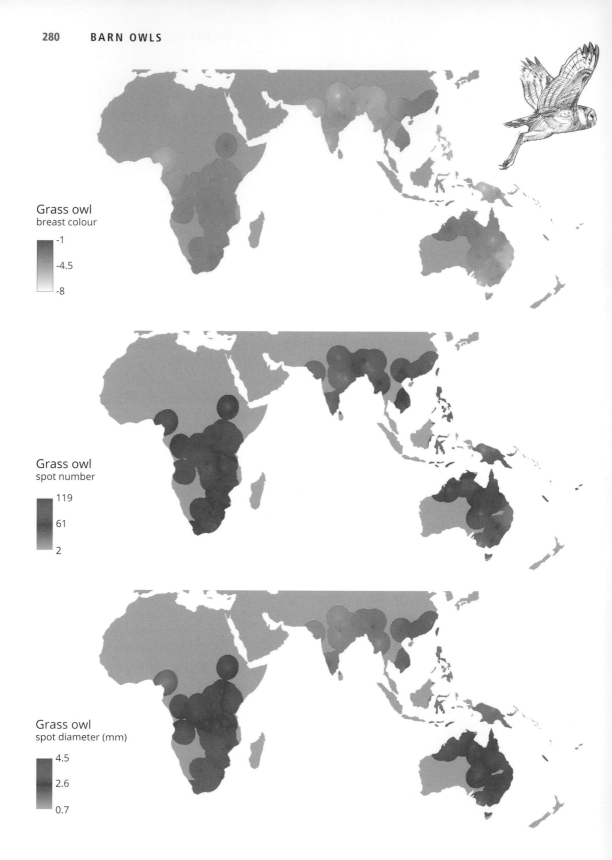

Geographic variation in the degree of reddish colouration, the number of black spots and the size of black spots in grass owls. For colouration the scale is from white (–8) to dark reddish (–1).

Masked owl
breast colour

-1
-4.5
-8

Masked owl
spot number

47
27
7

Masked owl
spot diameter (mm)

5.5
2.9
0.4

Geographic variation in the degree of reddish colouration, the number of black spots and the size of black spots in masked owls. For colouration the scale is from white (–8) to dark reddish (–1).

is six times stronger in Europe than in North America. In the southern hemisphere, the African grass owl, African and Australian barn owls and Australian masked owls follow the same geographic variation, with birds being more melanic (i.e. darker reddish and more spotted) in the temperate parts than in the equatorial parts of the continents. The opposite pattern, however, prevails in South America, with barn owls (*Tyto furcata tuidara*) being darker reddish and spottier closer to the south pole.

## Sex-specific clines
Plumage traits are sexually dimorphic, with females being, on average, darker reddish and spottier than males. These colour differences suggest (as discussed previously) that some plumage variants have a more important function in one sex than in the other, such as large black spots being favoured in females but counter-selected in males. If the benefits and disadvantages of displaying different plumage traits vary geographically, latitudinal variation could be stronger in one sex than in the other. Accordingly, males display larger spots in northern Europe than in southern Europe, whereas females have spots of a similar size across the continent. Perhaps in females selection to be large-spotted is strong throughout Europe, and in males only in northern Europe. Alternatively, the costs of displaying large spots in males may be less pronounced in northern Europe than in southern Europe.

The situation is fundamentally different with respect to the white to reddish colouration of the ventral body. Regardless of sex, individuals are mostly dark reddish in northeastern Europe and whitish in southern Europe. This observation could suggest that plumage reddishness has a similar function in both sexes, and that the costs and benefits of being dark reddish or whitish are similar in males and females across the continent.

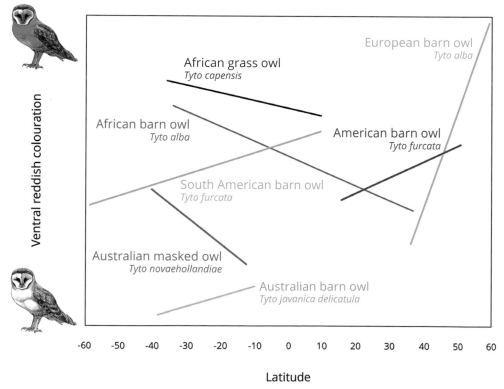

Geographic variation in the degree of reddish colouration, the number of black spots and the diameter of black spots located on the underparts of various Tytonidae. The regression lines indicate the strength of the relationship between latitude and plumage colouration. Latitudinal variation is particularly pronounced in Europe compared to other regions (*see also opposite page*).

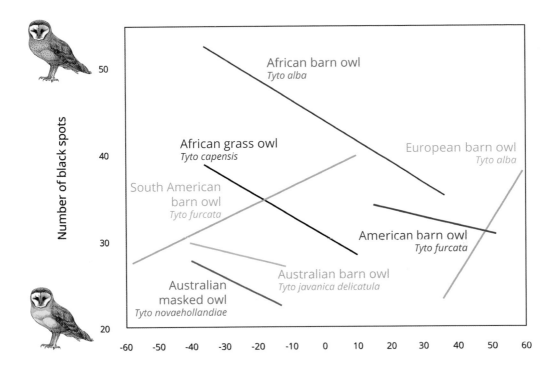

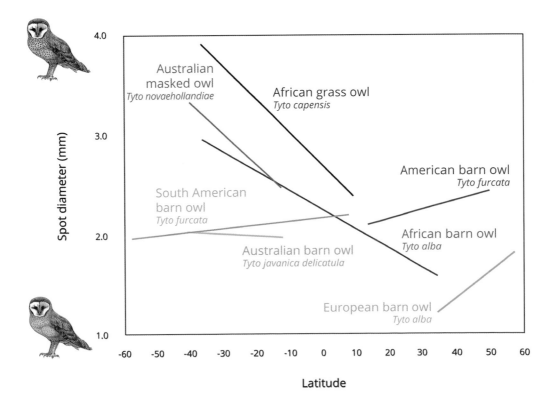

**Precipitation and temperature** Identifying which ecological factors triggered the evolution of latitudinal variation in plumage traits on different continents is not trivial. At such a large continental scale, most bio-climatic variables vary from north to south, emphasizing the difficulty in inferring to which environmental factors plumage variants are adapted in different populations. For instance, in the equatorial zone, there are more parasites, and the climate is less seasonal than in the temperate zone. Analyses were thus performed to examine whether ambient temperature and precipitation are associated with the evolution of plumage traits, as predicted by two biogeographic rules.

(1) In line with **Gloger's rule**, barn owls are more heavily pigmented in humid than dry environments. This is the case for the degree of reddish colouration at the global scale but only at a more local scale in relation to spot size, at least in the USA and the United Kingdom. Being darker in regions where precipitation is abundant might be an adaptation to protect the body against parasites that thrive in humid habitats. Additionally, in such luxurious habitats, light conditions are poor, and dark-coloured plumages could enhance camouflage. Finally, melanin could reduce feather wear caused by rubbing against dense vegetation.
(2) According to **Bogert's rule**, barn owls from the Americas, Africa and Europe evolved a darker reddish colouration in cooler regions and whitish plumage in warmer regions. Perhaps like cold-blooded organisms that need heat from the sun to accumulate warmth, a dark reddish colouration in the barn owl may have a thermoregulatory function. Interestingly, this relationship is not observed in Australasia.

The biological significance of the observations that barn owls exhibit smaller and fewer black spots as well as lighter reddish colouration on islands than on continents remains unclear. The same applies to the findings that on small islands barn owls are darker reddish than on larger islands, and that in the northern hemisphere barn owls display larger black spots closer to the equator. We still have a lot to learn about how plumage traits vary geographically!

## FUTURE RESEARCH
- The reasons why barn owls are darker reddish in humid and cold regions and whiter in dry and hot regions must be investigated.
- Why Bogert's rule applies to the Americas, Europe and Africa but not Australasia is unclear.
- Are geographic variations evolutionarily stable, or are they currently changing in relation to anthropogenically induced global environmental change?
- Why insular populations of barn owls are differently feathered compared to continental populations remains unknown.
- Did predator–prey interactions play a role in the evolution of plumage colouration at the global scale? Evaluating this possibility requires testing whether differences in diet between populations are associated with inter-population differences in colouration.

## FURTHER READING
Antoniazza, S., Ricardo, K., Neuenschwander, S., Burri, R., Gaigher, A., Roulin, A. and Goudet, J. 2014. Natural selection in a post-glacial range expansion: the case of the colour cline in the European barn owl. *Mol. Ecol.* **23**: 5508–5523.
Burri, R., Antoniazza, S., Gaigher, A., Ducrest, A.-L., Simon, C., The European Barn Owl Network, Fumagalli, L., Goudet, J. and Roulin, A. 2016. The genetic basis of color-related local adaptation in a ring-like colonization around the Mediterranean. *Evolution* **70**: 140–153.
Romano, A., Séchaud, R., Hirzel, A. H. and Roulin, A. 2019. Climate-driven convergent evolution of plumage colour in a cosmopolitan bird. *Global Ecol. Biogeogr.* **28**: 496–507.
Roulin, A. and Salamin, N. 2010. Insularity and the evolution of melanism, sexual dichromatism and body size in the worldwide-distributed barn owl. *J. Evol. Biol.* **23**: 925–934.
Roulin, A., Wink, M. and Salamin, N. 2009. Selection on a eumelanic ornament is stronger in the tropics than in temperate zones in the worldwide-distributed barn owl. *J. Evol. Biol.* **22**: 345–354.

Conclusion

# To the future

Over the years, the barn owl has proved to be an excellent model organism for the study of many key biological questions, spanning fields as diverse as physiology (vision, hearing, metabolism), ecology (predator–prey interactions, hunting behaviour), behavioural ecology (family relationships, dispersal), population dynamics, genetics and evolutionary biology (colour polymorphism). I have taken advantage of this huge body of knowledge to present this fantastic bird in a more comprehensive way than has been attempted in any previous book on the species, by considering all the aspects of evolutionary ecology that have been studied and doing my best to demonstrate the multitude of ways in which they are interconnected. This approach clearly shows that even when a researcher's desire is to understand a particular aspect of a species' biology, a thorough knowledge of all biological facts is necessary to fully appreciate the importance of the scientific question to be tackled.

While amateur ornithologists have always found the barn owl and its allies fascinating, scientists from the academic world initially thought that this group of birds was not worth the investment of so much time (we have to work by day and by night to study this bird) or travel (we have to cover many kilometres to collect data over very large study areas). I remember one professor asking me, 'Alex, why don't you work with snails rather than with barn owls?'

In the event, the barn owl has proved instrumental not only in providing several missing pieces of knowledge that were difficult to investigate in other birds, but also in casting a refreshing new light on old biological questions. The excitement of discovery motivated researchers to consider small details such as plumage traits that turned out to yield the most surprising results. How could we have predicted that white plumage helps the barn owl catch rodents, particularly under a full moon? And how could we have imagined in the 1990s that variation in the size of black feather spots could predict aspects of genetic quality? These results raised new paradigms of general interest far beyond the barn owl – for instance, an understanding of how different colour morphs can coexist and why and how melanin-based traits can be associated with physiology and personality. Because plumage traits are made of melanin pigments, just like variations in the skin of humans, studying the barn owl has important implications for understanding our own biology and history. We even recently started biomedical studies based on the finding that the *POMC* gene in the barn owl comprises a very strange genetic sequence. Studying the barn owl therefore helps us to understand the natural world in general.

Another key element of barn owl biology is seen in the intricate interactions that take place between family members. Although I always heard nestlings calling all night long even in the absence of their parents, I realized in 1997 that this behaviour made no sense according to the prevailing theories of parent–offspring conflict and sibling competition. Studying this behaviour led to the development of the 'sibling negotiation theory' that describes the way that young siblings communicate vocally to resolve conflicts peacefully. This is not only of interest from a pure scientific point of view, but it is also highly symbolic in the sense that a raptor with an array of weapons (sharp bills and claws) at its disposal is sufficiently wise to solve conflicts by negotiating rather than by fighting. This should be an inspiration for us humans.

We know a lot about the evolutionary ecology of the barn owl – though much less about its relatives, the other members of the family Tytonidae. The knowledge summarized in this book is based on an impressive number of scientific papers (well over 3600). This incredible volume of knowledge has not reduced the need for further research, however. On the contrary, it has opened the door to several further avenues of research, and has spurred ornithologists on to embark on many new projects, as the following four priorities for actions suggest.

(1)  **Fill the gaps in our knowledge.** A number of aspects of barn owl biology have not been considered in this book because of a lack of detailed studies. For instance, although the moult pattern has sometimes been described, the causes and consequences of different moult patterns

are unknown. Given the very long breeding season of the barn owl, breeders often replace wing and tail feathers during the incubation and rearing periods, which is relatively rare in birds. Other issues that still require research include the barn owl's sensitivity to various anthropogenic activities (poisons, noise, the impact of traffic on foraging, etc.) and population genetics/genomics at the global scale.

(2)  **Tackle new research questions.** This book has discussed a number of fascinating ideas that are of general interest beyond the barn owl. For instance, melanin-based colour traits are associated with many other phenotypes, such as resistance to numerous stress factors. From a functional point of view, why does this happen, and from a mechanistic perspective, how is it possible? These are important issues, given the presence of melanin pigments in most animals. Additionally, the cosmopolitan distribution of the barn owl suggests that its body plan is adaptable to a variety of different environments. What are the exact characteristics that make the barn owl so successful? The global distribution of the barn owl offers opportunities to examine not only the impact of key environmental factors (e.g. rainforest, desert, temperate climate) on the evolution of diversity of life forms but also the similarities and differences between barn owl populations on different continents. There is a growing need for studies to be undertaken outside Europe and North America. Additionally, the remarkable cooperative behaviour observed in young barn owl siblings deserves more research attention. Finally, the development of new technologies to follow owls through time and space with GPS should provide new insights into the life of a nocturnal predator, which is otherwise very difficult to observe.

(3)  **Spread the word.** Even though many key aspects of barn owl biology, such as diet and reproductive success, have been extensively studied, many ornithologists continue to collect data on those parameters. It is very important to continue gathering such data, so that we can evaluate whether populations are stable or are suffering from human activities at a local or regional level, or from global effects such as climate change. However, there is a need to increase the dissemination of findings. Most of these potentially valuable observations are still unpublished.

(4)  **Engage with the lay public.** The barn owl lives close to humans, and although previously believed to be a harbinger of doom, it has become a symbol of reconciliation and peace. In the Middle East, the barn owl has brought people from communities in conflict around the same table to discuss the bird's potential as a pest control agent, to replace rodenticides as a means of killing the rodents that are presently devastating agricultural fields in the region. Such stories should be widely disseminated beyond the scientific community. This development has been a bottom-up process spurred on by actors working to resolve cross-border issues. The programme demonstrates that people from diverse cultural, social and political backgrounds can successfully engage in dialogue when affected by the same issues. Using the barn owl as a symbol of peace is remarkable, considering that it was and still is perceived as a bad omen in many cultures.

I hope you have enjoyed reading this book as much as Laurent Willenegger and I have enjoyed working on it together. But my main wish is that readers will have learned a bit about the barn owl and will look at the species with new interest, inspired both to protect it (and nature in general) and to carry out research on it – or indeed on any other animal. Writing (or reading) a thick book about an organism might give the impression that our understanding of this organism is complete – but on the contrary, it rather shows that our interest is undimmed, and that we remain dedicated to the quest of seeking answers to numerous unresolved scientific issues.

# Species names

Scientific names of barn owls and their relatives (family Tytonidae) are given within the text. The names of other species mentioned are listed here.

Adelie penguin *Pygoscelis adeliae*
African eagle owl *Bubo ascalaphus*
African marsh harrier *Circus ranivorus*
American kestrel *Falco sparverius*
Aquatic warbler *Acrocephalus paludicola*
Azara's akodont *Akodon azarae*
Barn swallow *Hirundo rustica*
Barred owl *Strix varia*
Beech marten *Martes foina*
Black-backed jackal *Canis mesomelas*
Brown falcon *Falco berigora*
Bulwer's petrel *Bulweria bulwerii*
Bunny rat *Reithrodon auritus*
Carrion crow *Corvus corone*
Chimpanzee *Pan troglodytes*
Common buzzard *Buteo buteo*
Common vole *Microtus arvalis*
Darwin's leaf-eared mouse *Phyllotis darwini*
Eagle owl *Bubo bubo*
Eastern pygmy possum *Cercartetus nanus*
Etruscan shrew *Suncus etruscus*
European kestrel *Falco tinnunculus*
European siskin *Spinus spinus*
Feral pigeon *Columba livia domestica*
Ferruginous hawk *Buteo regalis*
Field vole *Microtus agrestis*
Flavescent colilargo *Oligoryzomys flavescens*
Giant eagle owl *Bubo lacteus*
Golden eagle *Aquila chrysaetos*
Goshawk *Accipiter gentilis*
Great horned owl *Bubo virginianus*
Great tit *Parus major*
Hamerkop *Scopus umbretta*
Hen harrier *Circus cyaneus*
Hobby *Falco subbuteo*
Hornet *Vespa crabro*
House martin *Delichon urbicum*
House mouse *Mus musculus*
House sparrow *Passer domesticus*
Indian flying fox *Pteropus giganteus*
Jackdaw *Coloeus monedula*
Lanner falcon *Falco biarmicus*
Lion *Panthera leo*

Little owl *Athene noctua*
Long-eared owl *Asio otus*
Long-tailed colilargo *Oligoryzomys longicaudatus*
Martial eagle *Polemaetus bellicosus*
Naked-rumped tomb bat *Taphozous nudiventris*
Osprey *Pandion haliaetus*
Peregrine falcon *Falco peregrinus*
Red kite *Milvus milvus*
Red-tailed hawk *Buteo jamaicensis*
Rough-legged buzzard *Buteo lagopus*
Ruddy shelduck *Tadorna ferruginea*
Ruff *Calidris pugnax*
Screech owl *Megascops asio*
Scops owl *Otus scops*
Shag *Phalacrocorax aristotelis*
Short-eared owl *Asio flammeus*
Small vesper mouse *Calomys laucha*
Snow goose *Anser caerulescens*
Social weaver *Philetairus socius*
Sparrowhawk *Accipiter nisus*
Swamp harrier *Circus approximans*
Tawny owl *Strix aluco*
Tengmalm's owl *Aegolius funereus*
Wild boar *Sus scrofa*
Wolf *Canis lupus*
Woodpigeon *Columba palumbus*

# INDEX

adaptive plumage traits
  colour, 264–267, 279
  number of spots, 275–277, 279
  size of spots, 262, 268–274, 279
adoption, 197, 198
  *see also* nest switching
age structure, 231
aggression towards humans, 63
agriculture, 21–22, 26, 30–31, 37–38, 40, 45,
    104–105
albinism, 241
allopreening
  adult, 219
  anti-parasite, 58, 219, 220
  exchange of commodities, 220
  massage, 220
  nestlings, 3, 200, 216, 219, 220
  reciprocation, 219, 220
  timing, 200
altruism, 217, 266
anthropocentrism, 28–29
anthropogenic environments, 21–22, 23, 26
  *see also* human structures
anti-parasite adaptations, 56, 58, 59
anti-predator behaviour, 63–64, 270
assortative pairing, 258, 259
asynchrony, *see* hatching asynchrony
autosome, 247, 253

begging behaviour, 185–187, 190, 203–209,
    210–214
  escalating, 211
  hunger level, 188, 211, 218
  parental response, 184, 185, 187, 189, 190,
    210, 211
  risk of predation, 62
  sibling competition, 203, 206, 211
  sibling negotiation, 204, 206, 210, 211
  towards parents, 206, 210, 211
Bering land bridge, 13
bill
  anti-parasite adaptations, 56, 58, 59

  heritability, 86
  length, 83, 85, 89
biological pest control, 30–31, 40–41
body mass, adult
  adaptation to flight, 75, 80, 93, 107,
    109, 264
  age, 193
  breeding failure, 34
  carrying prey, 75, 93
  food deprivation, 77, 80
  incubation, 156
  manoeuvrability, 75
  melanin-based plumage trait, 270
  pellet production, 74, 75
  reproduction, 191–193
  sexual dimorphism, *see* sexual
    dimorphism
  wing loading, 107, 109
body mass, nestling
  food supply, 174
  growth, 166, 173–175
  maternal nest desertion, 180
  melanin-based plumage traits, 258,
    270, 271
  overshoot, 173, 174
  rain, 174
  recession, 173, 174
  second brood, 177
  sexual dimorphism, 90
body size, 82–87
  effect of parental feeding rate, 86
  foraging, 89
  geographic variation, 82
  *see also* body mass
body temperature, *see* temperature
Bogert's rule, 284
breeding age, 131, 231, 235
breeding dispersal, 228–229
breeding failure, 142, 170–172, 180
breeding season, 152–154
  *see also* phenology
brood patch, 162

brood reduction, 3, 155, 163, 165, 166
brood size, 77, 165, 167, 170, 180, 184, 185, 246
brooding, 180, 221, 272
buildings, 23–26, 145–149

cannibalism, 215, 216
*Carnus hemapterus*, 59, 60, 270
caves
    nesting, 144, 145, 149
    roosting, 95, 97, 98, 124, 126
circadian rhythm, *see* diurnal activity, nocturnal
        activity
cliffs
    nesting, 23, 144, 145, 148
    roosting, 97
clutch size, 129, 142, 158 159, 160, 167
cold weather, 76–80
    *see also* winter
colonization, 19
colour polymorphism, 240–244, 264–267
    genetics, 245–251
communal breeding, 133–134
communal roosting, 99
competition, *see* sibling competition
consanguinity, 135
conservation, 27–52
cooperation between siblings, 3, 180, 203, 204
cooperative breeding, 133, 135
copulation, 129–131, 189
    clutch size, 131
    colour polymorphism, 244
    deserting female, 182
    extra-pair, *see* extra-pair copulation
    frequency, 130–132, 157
    timing, 128, 130
corticosterone, *see* glucocorticoids
cosmopolitan distribution, 2, 3, 18–22, 278, 287
courtship, 128–130
    winter, 128, 129
courtship feeding, 129, 178
cultural views of barn owls, 36, 49, 52

daytime activity, *see* diurnal activity
decapitation of prey, 189
desertion, *see* nest desertion
developmental homeostasis, 270
diet, 71–73, 118–126
    barn owl species complex, 118–125
    opportunism, 21, 116–117

prey selection, 116–117
    Tytonidae, 126
digestion, 215
    age, 73
    digestive juice, 73
    food deprivation, 77
    food intake, 196
    gastrointestinal tract, 73, 74,
    parasites, 56
    pellets, 73–75
    poor digestion, 72–75
disassortative pairing, *see* assortative
        pairing
diseases, 54–57, 58
dispersal, *see* breeding dispersal, natal
        dispersal
diurnal activity
    hunting, 63, 70, 93
    nestlings, 200–202
    roosting, 95–98
divorce, 138, 140, 229
double brooding, *see* second broods

ear, 66, 67, 69
ectoparasites, 58, 60, 61, 219, 220, 270
education, 49, 52, 287
eggs
    fertilization, 130, 135, 163, 164
    formation, 129, 154–158, 161
    size, 89, 155, 157, 158, 160, 256, 272
emigration, 231, 236
endoparasites, 54, 55, 58
erythrism, 241
ethics, 32–34
eumelanin, 240, 255
evolution, 10–15
    Africa, 13
    Americas, 14
    Asia, 13–14
    Europe, 13
    Oceania, 14–15
experimental design, 33–35
extra-pair copulation, 130, 133, 135, 181, 257
extra-pair paternity, 135
eye, *see* vision

facial disc, 21, 66–67
falls from the nest, 46, 49, 56, 190
farming, *see* agriculture

fasting
  daily fast, 93, 192, 215, 271
  food depletion, 77, 80
feather structure, 107–111
  barb, 108
  barbule, 108
  calamus, 107, 272
  rachis, 107
  serrations, 108, 109
  shaft, 107, 272
  transverse bars, 109, 110, 272
  vane, 107–109, 272
feather wear, 255, 272, 284
feeding rate
  father, 185
  mother, 171
  parents, 174, 184, 185, 194, 210, 211
fertilization, 130, 135, 163, 164
fidelity, 138–142
  see also divorce, nest-site fidelity
flight
  mechanics, 106–111
  silent, 3, 21, 108, 109, 111
  speed, 66, 73, 101, 106–109, 112, 116
  wing loading, 109
food
  allocation, 211, 214
  sharing, 3, 217–220, 266
  theft, 198, 211, 215, 216, 218
food requirement
  adult, 71–73, 93
  maternal plumage, 270
  nestling, 174, 196
food stores, 93, 194–196
  consumption, 192, 195
  determinants, 194, 196
  effect on reproduction, 194
  feeding offspring, 185
  outside breeding season, 128
  parental feeding rate, 185, 194
  sharing, 218
  theft, 93
foot, anti-parasite adaptations, 59
frequency-dependent selection,
     243, 244

gastrointestinal tract, see digestion
genetics
  genetic conflict, 249, 261, 262

good genes, 257
  plumage traits, 247, 248
Gloger's rule, 284
glucocorticoids
  allopreening, 220
  colour polymorphism, 244, 270,
     272
  oxidative stress, 269
  sibling negotiation, 206
  stress, 269

habitat, 102–105
habitat loss, 37–38
hatching, 161–162, 163–166
  before hatching, 163
  egg size, 155, 157
  maternal care, 162, 180, 185, 189,
     210
  mother–father relationship, 131
  parasites, 59, 60
  success, 161–164, 167, 238
hatching asynchrony, 3, 164–166
  age difference, 163
  allopreening, 220
  brood reduction, 3, 163, 165, 166
  comparison with other species, 165
  food sharing, 217, 218, 220
  incubation, 162
  parental care, 166, 192
  sibling competition, 86, 163, 165,
     166, 180, 206, 211, 214, 215,
     218, 222, 226
  thermoregulation, 221
hatching synchrony, 164, 165
hearing, 3, 21, 66, 68, 69, 112, 209
  see also ear, facial disc
home range, 100–105
hovering, see hunting methods
huddling, 180, 221, 222, 275, 276
human attitudes, 28–31, 46–52
human habitats, 21–22, 23, 26
human impacts, 39–43, 49
human structures, 145–149
  see also buildings, nest boxes
hunting methods, 112–115
  from a perch, 73, 106, 112, 113, 115
  hovering, 112
  on the wing, 21, 106, 108, 112, 113
  timing, 185

immigration
   impact on phenotypic adaptation, 21, 80
   population dynamics, 46, 139, 236, 244
inbreeding, 137, 227
incest, 135
incubation, 132, 161, 162
   brood patch, 162
   chittering calls, 210
   effort, 63, 89, 162, 164, 185, 191, 192
   failure, 35, 167, 171
   feeding, 129, 162, 185, 188
   hatching asynchrony, 164, 165
   hatching rank, 165
   kestrel eggs, 133, 198
   testosterone, 272
   time, 161
independence from parents, 104, 155, 173, 178,
      179, 181, 198, 225, 230
infanticide, 194, 215
infertility, 163
intestine, 73
   see also digestion
island syndrome, 86
islands
   colouration, 6, 253, 284
   diet, 41, 86, 119, 124
   dispersal, 13, 86, 227
   distribution, 13
   impact of barn owls, 40–41, 43
   introduction, 14, 41
   roost, 145
   speciation, 14, 15
Israel, see Middle East peace process

Jordan, see Middle East peace process

kin selection, 188, 217, 218

laboratory animals, 6–7
laterality, 270
laying date, see breeding season
leucism, 241
lice, 58–59
lift, 107, 109
literature, scientific, 8–9

manoeuvrability, 75
mate choice, 247, 257–259

mate guarding, 130
maternal effect
   begging for food, 189
   egg, 59
   maternally inherited genes, 138
mating system, 132, 133
melanin, 240, 247, 272, 273, 284, 286, 287
melanism, 241, 278
melanocortin-1-receptor (mc1r) gene,
      250
melanogenesis, 240, 247, 249, 271
Middle East peace process, 30–31
migration, 15, 22, 229
model system, 6–7
monogamy, 130, 135, 142
mortality, 230–234
   conservation, 44, 49
   food deprivation, 270
   population dynamics, 235, 236
   roads, see road deaths
   shooting, 234, 234
   size of plumage spots, 262
   starvation, 49
   temporal change, 230
   wind turbines, 234
   winter, 76
mutual preening, see allopreening
myths, 36, 44, 49, 52

natal dispersal, 137, 224, 225–227, 228,
      236, 266
   Europe, 225
   North America, 226
negotiation, see sibling negotiation
nest boxes, 45–47, 148–149, 151
   monitoring, 42
nest desertion, 178–180
   changing site, 181
   by female, 182
   by male, 179, 181
   effect of brood size, 178, 180
   effect of date, 180
   female quality, 180
   fitness effects, 171, 180
   frequency, 179, 180
   reproductive success, 180
   second brood, 179, 180
nest-site fidelity, 138, 140, 178, 228, 229

nest sites, 23–26, 144–151
  artificial, 145–149, *see also* nest boxes
  competition, 150–151
  ground, 13, 26, 63, 144
  hygiene, 56
  trees, 23, 26, 145, 148, 149
  underground, 144, 145
nest switching, 198
  *see also* adoption
nestling growth, 59, 73, 86, 155, 157, 165,
    173–175, 270, 271
nocturnal activity, 92–94
  anatomy and physiology, 68–69
  behaviour, 69–70
  hunting, 92–93
  visual acuity, 69

opportunistic diet, 21, 116, 117
oxidative stress, 269, 270

Palestine, *see* Middle East peace process
parasite resistance, 240, 268, 270, 273, 278, 284
parasites, 54–61
  *see also* ectoparasites, endoparasites
parent–offspring conflict, 286
parental roles, 184–190
paternal care, 129, 185, 188, 196
peace, *see* Middle East peace process
pellets, 73–75
  content, 2, 30, 73, 74, 118
  ejection, 75
  hygiene, 56
  nest material, 149
  number of bones, 72, 73
  number per day, 75
  pollutants, 39
  sodium content, 73
  sub-fossil pellets, 124, 126
  suffocation, 75
persecution, 36
personality, 241, 268, 286
pest control, biological, 30–31, 40–41
pesticides, 26, 30, 39, 40, 235, 287
pet trade, 49, 233
phaeomelanin, 240, 255
phenology
  determinants, 153, 154
  variation, 152, 154, 156, 157, 165

*Phodilus* (bay owl), 13, 105, 158
phylogeny, 11, 14–15
*Plasmodium*, 54
pleiotropy, 240
plumage traits, 240–284
  adaptive, *see* adaptive plumage traits
  age, 254–256
  alternative strategies, 3, 243, 244, 266
  colour, 240–244, 279
  dispersal, 264, 266
  genetic correlation between sexes, 248, 249
  genetics, 245–251
  geographic variation, 278–284
  grass owls, 280
  heritability, 247
  masked owls, 281
  mate choice, 257–259
  number of spots, 275–277, 279
  parasite resistance, 240, 268, 270, 273,
    278, 284
  personality, 241, 268, 286
  sex, 248–249, 252–253, 260–263, 282
  size of spots, 262, 268–274, 279
  Tytonidae, 282–283
pollution, 39–41
  *see also* pesticides
polyandry, 88, 133, 178
polygyny, 132, 133
polymorphism, *see* colour polymorphism,
    plumage traits, sexual dimorphism
population decline, 36–38
population dynamics, 3, 235–238
  *see also* immigration, mortality
predation, 270
  on barn owls, 62–64
preening
  ectoparasites, 58
  mother, 266
  nestling, 200, 220, 270, 272
  *see also* allopreening
prey, *see* diet

reciprocation, 218, 219
reconciliation, *see* Middle East peace process
rehabilitation, 46, 49
replacement clutches, 152, 167, 171, 172,
    180, 194
reproductive potential, 3, 21

reproductive season, *see* breeding season
research, 8–9, 32–35, 286–287
reversed sexual size dimorphism, *see* sexual
    dimorphism
road deaths, 26, 36, 44, 149, 230, 231, 232, 235
rodenticides, *see* pesticides
rodents as prey, 118–126
roosting, 95–99
    buildings, 26, 95, 98, 145, 148
    communal, 99
    foliage, 97
    foraging, 100, 113, 115
    forest, 97, 105
    frequency of use, 97, 98
    ground, 97, 99
    height above ground, 97
    nest, 98, 99, 129, 173
    nestlings, 174, 197
    rocky sites, 97, 124
    roost choice, 99
    sex differences, 97, 129
    sooty owl and masked owls, 95, 97
    trees, 95, 97, 145
    underground, 97
    winter, 194
    *see also* sleep

scientific studies, 8–9, 32–35, 286–287
second broods, 176–178
    clutch size, 158, 159, 176
    copulation, 131, 178, 182
    determinants, 152, 177, 178
    dispersal, 227
    frequency, 172, 176, 177, 180
    interaction with first brood, 174
    laying date, 159, 176, 178
    mate change, 181
    maternal desertion, 171, 178, 179, 180,
        181, 218
    nest site, 46, 178
    parental investment, 177, 178
    success, 176, 177
    timing, 158
senescence, 238, 238
sex chromosomes, 247, 248, 253
sexual behaviour, 128–142
sexual dimorphism, 88–90, 250, 252, 253,
    262, 282
sexual maturity, 131, 235

sexually antagonistic selection, 260–262
Shannon's diversity index, 119–121
sharing of food, 217–220, 266
shrews as prey, 122
siblicide, 215, 216
sibling competition, 3, 204, 211, 286
sibling interactions, 200–222
sibling negotiation, 200, 203–209, 210, 211, 215,
    219, 286
silent flight, 21, 108, 109, 111
sleep, 93–94, 113, 150, 173, 200, 202, 275, 276
    electroencephalography, 93, 202
    non-REM sleep, 202, 276
    rapid eye movement sleep (REM), 94, 200, 202
    *see also* roosting
snow, 76–77, 153, 155, 230
soliciting behaviour, *see* begging behaviour
speciation, 14, 15, 21
spots in plumage
    number, 275–277, 279
    size, 262, 268–274, 279
starvation
    brood reduction, 157, 165, 216
    cause of mortality, 49, 76
    death of father, 171
    falling out of the nest, 49
    food depletion, 77, *see also* fasting
    sex differences, 80
stomach
    digestion, 73
    foraging, 93, 109, 264
*Strigea*, 56
survival, *see* mortality
synchrony, *see* hatching synchrony

tail
    heritability, 86
    number of feathers, 87
    rigidity, 272
temperature
    ambient temperature, 46, 71, 145, 149, 152,
        164, 167, 170, 221, 222, 230, 275, 276, 284
    body temperature, 76, 77, 132, 170, 210, 221,
        222, 275
    thermoneutral zone, 77, 221, 222
territorial behaviour, 46, 100, 104, 138, 140, 181,
    224–226, 228, 229, 266
testosterone, 271, 272
theft of food, 198, 211, 215, 216, 218

thermoregulation, 221
threats, *see* human impacts, persecution, pesticides, pollution
ticks, 58–60
*Tyto alba* subspecies, 13, 105
*Tyto almae* (Seram masked owl), 14, 105
*Tyto aurantia* (golden masked owl), 14, 105
*Tyto furcata* (American barn owl), 15, 282
*Tyto gigantea*, 10
*Tyto glaucops* (ashy-faced owl), 13, 15
*Tyto inexspectata* (Minahassa masked owl), 14, 105, 149
*Tyto javanica delicatula* (Australian barn owl), 14
*Tyto javanica javanica* (Asian barn owl), 14
*Tyto javanica stertens* (Indian barn owl), 13
*Tyto multipunctata* (lesser sooty owl), 15, 105
*Tyto nigrobrunnea* (Taliabu masked owl), 13, 14
*Tyto rosenbergii* (Sulawesi barn owl), 13, 97, 105, 149
*Tyto sororcula* (Moluccan masked owl), 14
*Tyto soumagnei* (Madagascar red owl), 13, 105
*Tyto tenebricosa* (sooty owl), 15, 105
*Tyto thomensis* (São Tomé barn owl), 13
Tytonidae
  evolution, *see* evolution
  cosmopolitan distribution, *see* cosmopolitan distribution

uneaten food, 195–196
  *see also* food stores
uropygial gland, 272

vision, 66, 68, 70, 112
  acuity, 68, 69
  eye, 68, 69, 113
vocal negotiation, 205–207
voles as prey, 122–123

wing
  heritability, 86
  island syndrome, 86
  length, 84, 85
  rigidity, 272
wing loading, 107, 109
winter
  adaptation, 80
  courtship, 128, 129
  fat accumulation, 76, 77
  foraging, 93, 113, 159
  home range, 104
  migration, 22, 77, 228
  reproduction, 77, 145, 152, 153, 155, 159, 164
  roost, 194
  *see also* cold weather, snow
winter mortality, 44, 76, 80, 188, 230, 232, 235, 237